250 Home Plans

FOR TODAY'S LIVING

FROM THE DRAWING BOARD OF

William G. Chirgotis
ARCHITECT

 Federal Marketing Corp.

DEDICATION

This book is dedicated
to Phil Birnbaum, founder of
Federal Marketing Corporation

CONTENTS

WHY AN ARCHITECT DESIGNED HOME?

Because an architect is seldom consulted by the average family, his is a little-understood profession. Most people think of an architect as a man who designs skyscrapers, factories, churches, schools and other massive structures that make up a crowded community. Few realize that the majority of architects contribute their talents to solving today's rural and suburban residential problems—the problems that the average family faces when they plan to have a new home. So few families realize this fact, that less than 5% of them start their planning by visiting an architect. The other 95% visit a model home built in a development, or perhaps buy blueprints from a stock plan service. Either way, they may still be receiving indirectly the benefit of architectural talent, if the model or the stock plan represents the work of an architect. A house designed by an architect will certainly give the potential home owner a correct layout, proper design, and good dollar value.

If you intend to use stock plans to construct your home, be sure that they are the work of a reputable architect. By making the blueprints of these homes available at a small fraction of their original cost to prepare, thousands of families that build from our plans enjoy a better standard of living. Avoid plans that bear the name of a designer but not an architect. The largest investment of your lifetime deserves the insurance of an architect's name indicating the authenticity of the design.

In many sections of the country there are no architects for miles around. Even in outlying sections of major cities, architects are often too busy to be available for small home work. If this is the case, you will undoubtedly consider purchasing a stock plan. A stock plan is a home design that is already in the form of working drawings. By permitting the resulting home design to be published, the architect makes the plan available to other families with similar requirements, and it costs them only a fraction of the original cost. They receive copies of substantially the identical blueprints from which the house was constructed, perhaps with some improvements and refinements gained from experience in building it. If you do employ an architect, one or more stock plans can serve as a starting point for your discussions, serving to crystallize your ideas and accelerate the planning. If the plan you buy is architect-designed, you can be certain that when it is properly executed by a competent builder, you are going to be the proud owner of a home that is solidly built, space engineered, comfort-endowed, and esthetically appealing. In short, you insure your investment by getting the maximum house per dollar spent. The homes illustrated in this book have layouts suited to a great segment of families all over the country and in popular income brackets. As you leaf through the portfolios that follow, you will see many homes that you feel you have seen and admired before. It is quite likely that you have driven past them and envied their lucky owners. They all exist. They are now available to you.

It is the aim of this book to display the home designs of one of this country's leading architects, William G. Chirgotis, to make these homes available to the home building public, and to point up how architectural services can help any family to attain a home that meets their highest aspirations.

A GUIDE TO ASSIST YOU IN SELECTING A HOME DESIGN TO FIT YOUR NEEDS!

Of the many homes illustrated in this plan book, one is certain to be your "Dream Home" meeting your budget and family requirements. The question is:—Which One?

The answer, of course, depends on a number of factors. Some are purely personal considerations of taste. Others are basic family needs. Still others are financial. It will be necessary for you to analyze each design on all counts. Here are some of the important points to keep in mind.

NUMBER OF ROOMS:

Requirements depend on the size of the family. If financially feasible, it is a good idea to provide separate bedrooms for each child and at least two bathrooms if there is more than one child. If there is a "third generation"—for example, a grandparent—a separate bedroom with private bath is virtually essential.

STYLE:

Colonial or contemporary? This question depends largely on personal preference or taste. It is unwise to build a home that is radically "different" from other homes in the neighborhood.

TYPE:

Each basic house type has its own definite advantages. The one-story ranch allows for easy living and maintenance. The two-story and the 1½ story Cape Cod offer low cost per square foot, also the complete separation of entertainment and sleeping areas. The split level and "multi-level" combine the best features of both types.

Ranch

An all-inclusive word that covers virtually any house in which all the rooms are on one floor at ground level. Because of the general truth that it costs more to build horizontally than vertically, the cost of a ranch, on the basis of the amount per square foot, is usually higher. The maintenance of a ranch is easier. Stair climbing is non-existent or minimal.

Multi-Level

Sometimes called by other names, such as Hi-ranch and Bi-level. In this type of house, the front foyer is at ground level, with a stairway upward to the main living area and another downward to what would ordinarily be the basement. Because the basement is raised out of the ground enough to permit windows above ground, the area is utilized for living purposes and usually contains a recreation or informal room.

Split Level

Has three or four levels. Less stair climbing when going from one level to another, but total climbing may be more than in a two-story. Especially suitable for rolling terrain. Lends itself to attractive exterior appearance if well designed. Requires more land than a two-story, but has more liveable space for the money than a ranch.

Two Story

Cost, on the basis of amount per square foot, is usually lower than other types of houses. Bedrooms have more privacy. Should have at least two bathrooms. More rooms can be built on less land. Many different architectural styles are available.

Expandable Ranch

The half-a-story usually refers to an attic which can be finished at the time of the original construction or later on. Often has master bedroom on first floor, children's bedrooms upstairs. Provides extra storage space under eaves. Has knee walls and sloping ceilings upstairs. Most one and one-half story houses have traditional details in the Cape Cod style.

Vacation and Leisure-Time Homes

Whatever your taste, whatever your budget, the following designs for vacation or leisure-time living offer a change from everyday patterns. Today — more than ever before, Americans are investing in the future in a "second" home — it pays dividends in pleasure and relaxation, while increasing in value over the years.

Whatever your choice, the following designs will intrigue your imagination and compliment your budget.

Dome Homes

Introduced about 20 years ago, this unique concept of living is today enjoying new phenomenal popularity.

Technically, the dome home originated from the sphere, nature's most favored and efficient means of enclosing unobstructed floor space economically.

Due to inflation and the continued ever increasing building construction costs and the fact that the factory assembled triangular space frames are simply bolted together on the site to form the finished building, drastic reductions are possible on quantities of building materials and on-site labor costs. As much as 20 percent less for a dome home than for conventional housing.

The dome provides a living area that also answers the need for efficient energy consumption and is particularly adaptable to solar heating.

Today's dome homes are attractive and offer an exciting new way of living. The minute you step inside the front entrance you are surrounded by fascinating forms and deceptively large spaces.

Because air naturally travels in a circular pattern, heating and/or cooling a dome home is more efficient and economical.

The dome provides maximum enclosed space with minimum surface area which means efficiency in terms of heat-gain or heat-loss. It has been estimated that heating and cooling costs can be reduced by at least 25 percent.

Americans are shedding the conservative trappings of their urban life and are adopting a more youthful, modern and exciting life style.

HOW TO READ A FLOOR PLAN

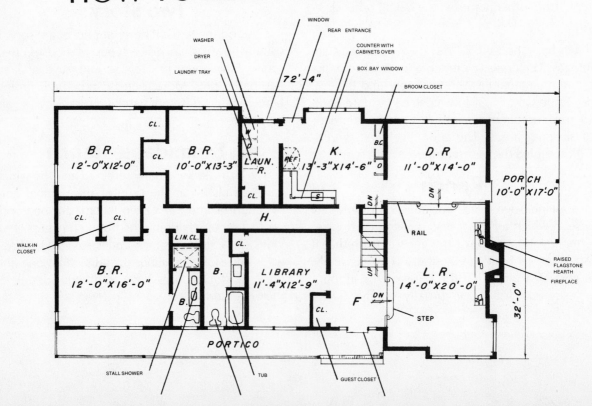

RANCH

An all-inclusive word that covers virtually any house in which all the rooms are on one floor at ground level. Because of the general truth that it costs more to build horizontally than vertically, the cost of a ranch, on the basis of the amount per square foot, is usually higher. The maintenance of a ranch is easier. Stair climbing is non-existent or minimal.

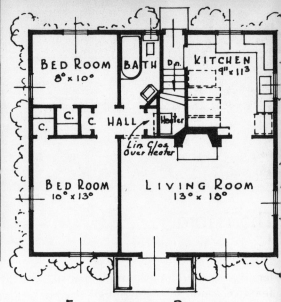

The Knox

Here is a compact home to fit even the most modest pocketbook; small enough to fit on a 40' lot. This plan provides for heat through a "closet heating unit" or if you wish, the boiler can be installed in a portion of the full basement.

The two comfortable bedrooms share a closet wall and each has two exposures. A cheery colonial fireplace adds to the charm of the living room. This compact plan makes housekeeping easy and combines all the features of a much larger home.

TOTAL LIVING AREA: 755 sq. ft.

FLOOR PLAN

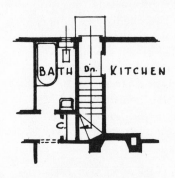

The Revere

A "Ranch" home seems to be the dream of many folks today. The name itself brings to mind the carefree and informal pattern of the West's wide open spaces. Of course, today, spaces are not quite so wide open in many areas, so a new ranch style has evolved, somewhat condensed for economy, but still containing the major essentials of the true ranch. This plan, as you will discover upon examination, brings to you very economically, dollar and space-wise, all the comfort and convenience of the ranch type.

TOTAL LIVING AREA: 1,207 sq. ft.

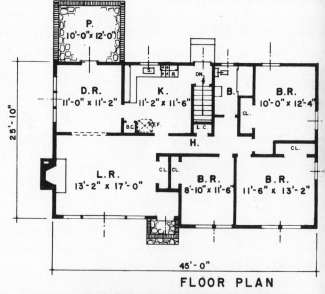

FLOOR PLAN

The Tilford

Long and low, this economical ranch home is stretched across the front for that "large-house" look. In spite of its length this is not an expensive house, yet contains all of the features found in many homes of much greater size and cost. A flagstone fireplace hearth extends to the front entrance to form a small entry area which includes a large guest closet. The living-dining ell is extra large and provides large wall areas for decoration and furniture arrangement. An ample kitchen provides dining space at the windows and has a separate small foyer at the service entrance which also leads down to the basement.

The three bedrooms are well separated from the living area and in addition to being served by one full sized bath, there is also a small private lavatory off the Master bedroom.

TOTAL LIVING AREA: 1,230 sq. ft. (excluding garage)

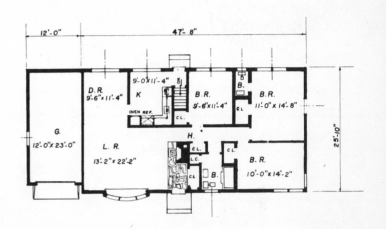

The Sutton

This three-bedroom ranch design derives its eye-catching character through its centrally located landscaped atrium, the three walls of which are almost completely glass and help bring an outdoor atmosphere into the house.

Inside,—a flagstone paved foyer effectively zones the activity areas;—the kitchen to the right, the dining-living area straight ahead and the small hallway around the perimeter of the atrium facilitates the flow of traffic to the other rooms.

If convenient and comfortable living is of primary importance in your search for a new contemporary ranch design with an outdoor atmosphere—this could be just the thing.

TOTAL LIVING AREA:	1,250 sq. ft.
Atrium	255 sq. ft.
Garage	529 sq. ft.

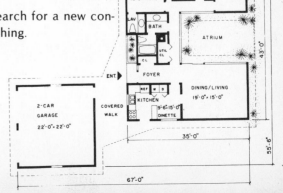

The Neptune

If you're just starting out in life, and can't afford a big new home—but still want to enjoy the very real benefits of home ownership and want to build a basic, comfortable house with the proper blend of economy and style—you would do well to consider this three bedroom design.

Inside, the architect has utilized every inch of the 1,275 square feet of living space; there seems to be a tremendous amount of habitable area in a house of this size, with everything needed by a family of two, three or four that wants all the rooms on one floor.

There is no doubt that the "all-on-one-floor" living of this clean and simple design is also especially appealing to retired couples or busy mothers of small families.

TOTAL LIVING AREA: 1,275 sq. ft.

The Fairfield

Here is a two-bedroom ranch house, planned for the luxury "L" that is so popular today. There's an air of informal elegance outside, with smart shingle and siding trim set off under a wide sheltering overhang. Note the handsome picture window overlooking the front lawns, and the simple uprights to give feeling at the front door. Inside, a vestibule, foyer and coat closet separate the living areas effectively. Just a few steps to the kitchen, a sunny, pleasant work area, and beyond there is the dining room, square and handsome, with picture window facing the back garden and a row of glass blocks in the wall opposite to add extra light. You can screen the dining room at dinner time or throw it open with the living room for more than 30' of hospitality space. One of the best features in a through living room like this one is that you can enjoy picture views, front and back, and the terrace opens out for gracious outdoor living in complete and sheltered privacy.

TOTAL LIVING AREA: 1,297 sq. ft.

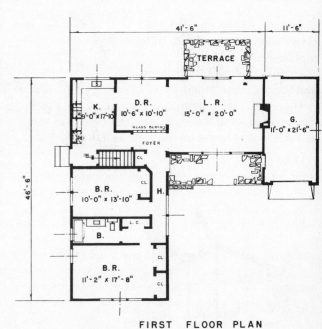

FIRST FLOOR PLAN

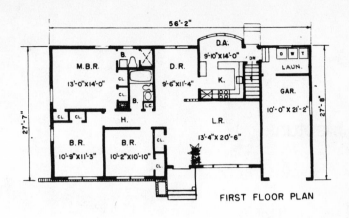

FIRST FLOOR PLAN

The Stanford

Families with an "in with the new, out with the old" philosophy of life will be attracted to this clean-lined contemporary three bedroom ranch design with its dramatic use of glass and unique roof lines.

The foyer is separated from the living room by a decorative wrought iron railing. This feature and the beamed cathedral ceiling of the living room visually enlarge this area.

This exquisite contemporary ranch design that is planned to achieve both economy and convenience and has all the ingredients of a larger home; it is ideal for a first home for young families or as a retirement home for mature couples.

TOTAL LIVING AREA: 1,306 sq. ft.
    Garage   254 sq. ft.

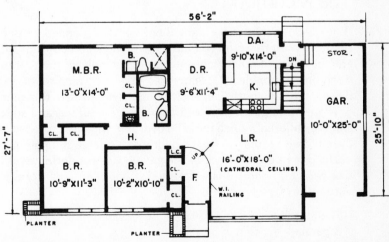

The Granada

The Spanish flavor of the old Southwest is delightfully captured and comes to life in the form of this enchanting ranch.

An interesting treatment of mixing rough stucco finish, projecting stained wood beams, arched picture window, low pitched roofs and stone-veneer, lends an exotic air of a Spanish villa to this latest design.

Inside,—the areas are planned for easy living, from the central entrance, you can reach any room with a minimum of steps from the small but adequate foyer which serves as an efficient traffic control center. A decorative wrought iron railing separates the foyer from the living room to the right. This feature and the beamed cathedral ceiling of the living room enlarge the entire area.

The bedroom area has three bedrooms clustered around a minumum hallway and is clearly delineated to maximum privacy and good sound conditioning with a buffer zone of closets and bathrooms.

This design modeled after the one-floor rambling structures built in the open spaces of the west can be a source of pride in any neighborhood.

TOTAL LIVING AREA:	1320 sq. ft.
Basement	1320 sq. ft.
Garage	550 sq. ft.

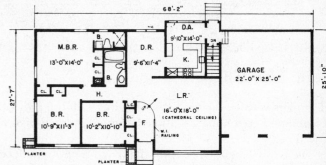

The Richmond

The ideal ranch plan for many is one which is planned around a center hall, with rooms easily accessible. With this plan you have the privacy and comfort of a 2 story house combined with the gracious ease of ranch house living. To the right of the entrance hall is a 20' living room opening freely in the popular "L" shape to a full-size dining room. Large window areas front and rear in these rooms add both light and through ventilation to this spacious entertaining area.

There's a large kitchen with ample work counter, and space enough for a table for breakfast and the children's lunch too. The bedroom wing is separated enough to lend privacy and quiet for the children's rooms, and each bedroom has large sliding-door closet space.

The open porch in back of the garage is a wonderful asset for summer evenings, and the oversize garage provides space for the car and plenty of storage for all garden tools and children's outdoor toys.

TOTAL LIVING AREA: 1,336 sq. ft. (excl. porch
   & garage)

The Kilmer

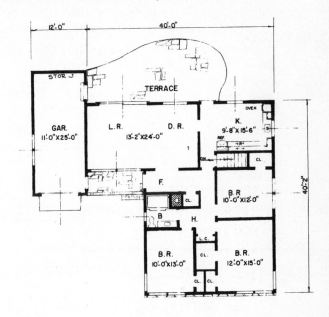

A refreshing exterior is boasted by this popular ranch design which will make it a hit in any neighborhood.

Including three airy bedrooms, a modern kitchen and the combination living-dining room, the interior comes to a graceful conclusion at a friendly terrace which is found at the rear of this breath taking home.

TOTAL LIVING AREA: 1,320 sq. ft.

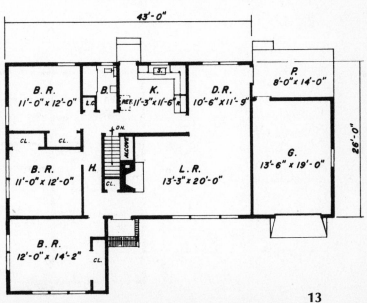

The Hilton

This plan presents a "different" arrangement for ranch living. The living room, with the narrow end facing front and containing a large picture window, has a fireplace on the porch side, backed up with a barbecue on the porch. The dining room projecting off to the side forms an "ell" for the porch which is open on two sides to catch the breeze. There is a large, yet stepsaving kitchen with practical circulation to all other parts of the house and to the outdoors.

Three bedrooms, all roomy, are grouped conveniently around the bath. The bath in itself is a feature—convenient to bedrooms, yet close to entry and arranged for use as a powder room for guests as well as normal facilities for the family.

TOTAL LIVING AREA: 1,365 sq. ft.

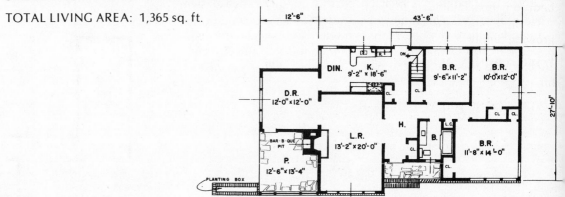

The Parkview

The massive chimney and long planting bed accent the appearance of this cozy three bedroom house. A large entrance foyer is featured here, flanked by living room and dining room and leading directly through the kitchen in the rear. Off to the left a hall leads to the bedroom wing affording maximum privacy for this area. There is a stair, conveniently located in the kitchen at the outside entrance, which leads to a fully excavated basement with unlimited space for recreation room, laundry area and workshop.

TOTAL LIVING AREA: 1,370 sq. ft.

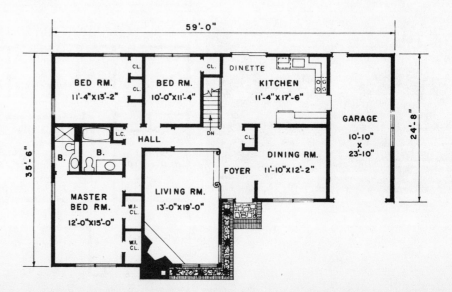

The Balmoral

Contemporary styling of this low ranch, with an exterior of vertical siding and brick veneer, combined with a low pitched angle roof extending over a conveniently located car port and exterior storage, portrays the "modern" living comfort incorporated into this plan.

The centrally located family room, opening directly to the kitchen, dining and exterior patio provides for ideal circulation and also provides for "inside-outside" living.

Note the built-in Bar-B-Que for lawn parties and cook outs; every couple's dream.

The utility room with incorporated laundry facilities, located adjacent to the bedroom area, minimizes laundry travel to and from the bedroom area.

Its three family size bedrooms satisfy the needs of a young growing family.

TOTAL LIVING AREA: 1,400 sq. ft. (carport excluded)

The Belmont

Here is a perfect picture of contentment as characterized by this homey ranch design.

A wrought iron column, colonial shutters, and a handy planting box each add to the beautiful exterior.

Inside we find a friendly living room and dining room convenient to the step-saving kitchen. To one side, three bedrooms are grouped together each with ample closet space. For the future, plans have been made for an upstairs bedroom and bath.

AREA: First floor   1,430 sq. ft.
         Second Floor Expansion   439 sq. ft.

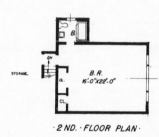

· 2ND · FLOOR PLAN ·

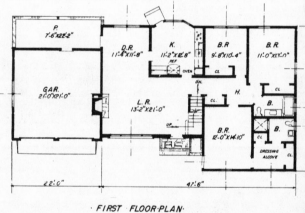

· FIRST FLOOR · PLAN ·

The Marlboro

Here is a modern ranch style home that has made a hit with so many prospective home owners.

There are three cozy, yet spacious bedrooms, each containing more than ample closet space. Convenient to your bedroom there are 1½ well designed baths.

The full size living room directly off the entry hall opens through an archway into a beautiful dining room. There is direct accessibility into the cheery and convenient kitchen from the dining room and from the entry hall.

A relaxing breeze-way and roomy garage round out this design for contented living.

TOTAL LIVING AREA: 1,434 sq. ft.

The Leeds

The shape of this modern design three bedroom ranch is in the form of an "L" which enables it to fit on a smaller lot than if all the rooms were placed within a conventional ranch outline.

It is enhanced by the horizontally-paned corner windows, stone veneer and the V-joint boarding in the gable that follows the roof pitch.

TOTAL LIVING AREA: 1,430 sq. ft.

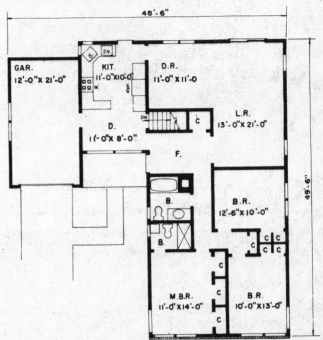

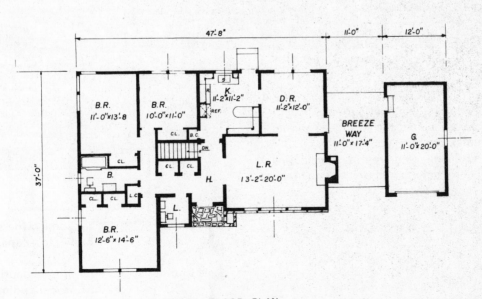

· FIRST FLOOR PLAN ·

17

The Glenview

The "Glenview" is a home for living, for entertaining and for comfort. It will delight in receiving your guests into its cheery spacious atmosphere. The large flagstone floored foyer with generous guest closet, short divider wall and attractive planter will usher you into an enormous area for living and dining. It is open from the front picture window to the rear glass wall and set off by the centrally located fireplace with its interesting free form hearth.

The kitchen will accept with pleasure your praises for its spaciousness; its multitude of cabinets and generous work top areas, plus the separate dinette area.

With cool quiet the three bedrooms will offer solitude and comfort for your sleeping hours; ample and convenient storage space for your wardrobe and modern efficiency, privacy and beauty for your toiletry in either of two full baths.

TOTAL LIVING AREA: 1,470 sq. ft. (excl. garage.)

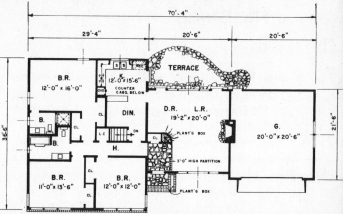

The Barton

Three ample bedrooms, each with roomy closets, and two full baths round out the secluded slumber area of this popular ranch design.

The living and dining rooms form an enchanting "L" which is ideal for entertaining and relaxing.

Adjacent to the complete kitchen we find a handy laundry to aid Mom in her washday chores.

TOTAL LIVING AREA: 1,485 sq. ft. (excluding porch)

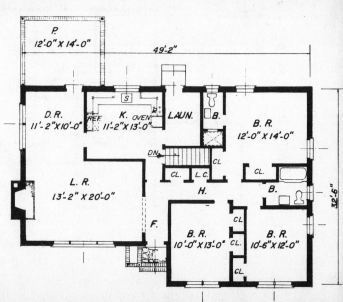

18

The Landis

The modern lines of contemporary architecture are dramatically expressed on the exterior of this three bedroom one-story design. Natural wood siding applied vertically on the exterior and with the windows facing the street-side held to a minimum, the floor plan inside offers informal living that many young families will find to their liking.

A sheltered entry leads to the spacious foyer, and the large glass panel over a low flower box affords a view of the rear scenery. Easy passage is available to the combination dropped living area and the family room and outside to the wrap-around flagstone patio through the sliding glass doors.

The bedroom wing is isolated from the center of activity and is well compacted into a convenient unit to the right of the entrance foyer.

If convenient and comfortable living is of primary importance in your search for a new home, this modern ranch-style design could be just the thing.

TOTAL LIVING AREA: 1468 sq. ft.
    Basement   790 sq. ft.
    Garage & Laundry   622 sq. ft.

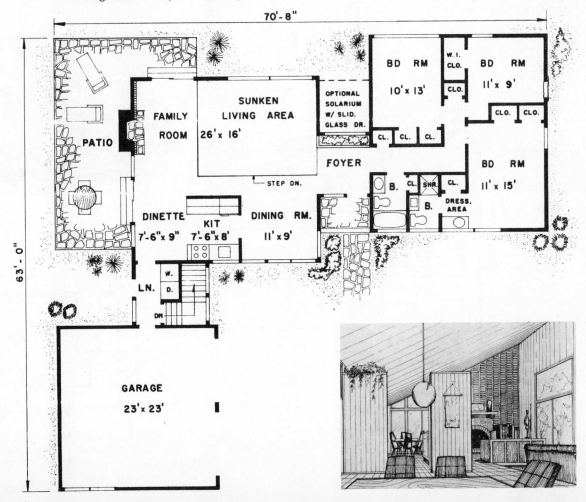

The Channing

In today's housing world, more and more families are looking for homes that are economical, yet distinctive and attractive enough to take real pride in ownership. This contemporary three bedroom ranch design, with shadow box windows and natural wood siding applied vertically on the exterior has a floor plan that many young families will find much to their liking.

The bedroom area has three bedrooms clustered around a minumum hallway and is clearly delineated for maximum privacy. A private bath services the master bedroom that has the quiet rear corner of the house. There are two closets, one of which is a walk-in, and a vanity in the dressing area. To the front are the other two bedrooms, for children or guests, with excellent wall space and double closets.

This contemporary ranch planned for both economy and convenience is ideal for a first home for young families or as a retirement home for mature couples.

TOTAL LIVING AREA: 1,483 sq. ft.
Garage: 550 sq. ft.

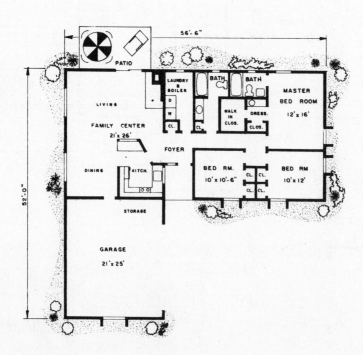

The Grandville

Generous glazing, a hallmark of contemporary styling, and interesting angles softened by the use of fieldstone veneer and vertical redwood siding give this three bedroom ranch a distinctive modern look.

From the entrance foyer you get an impression of roominess, for you can view the living-dining room as well as the family room with the patio beyond the sliding glass doors. Notable features are the dramatic corner fireplace with wraparound raised flagstone hearth and the sloped beamed ceilings in the living areas which add a sense of spaciousness.

The bedroom area has three bedrooms clustered around a minimum hallway and is clearly delineated for maximum privacy and good sound conditioning with a buffer zone of closets and bathrooms.

For active families whose style is casual, this single story contemporary design is an ideal choice.

TOTAL LIVING AREA: 1485 sq. ft.
     Laundry & Garage   625 sq. ft.
     Basement   1570 sq. ft.
     Patio   225 sq. ft.

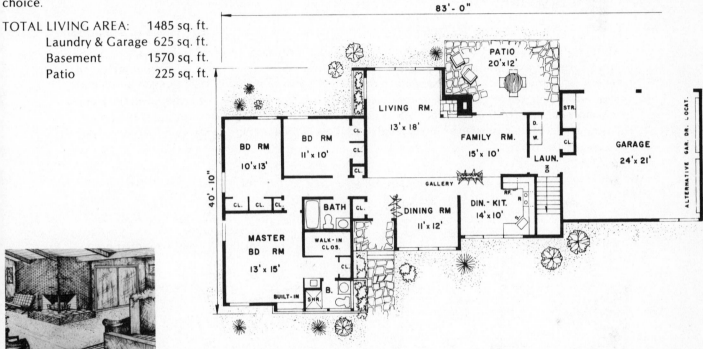

The Scarsdale

This particular design is best suited for the homeowner trying to capture the peace and beauty of nature in multi-level living, in the contemporary manner. An exterior of vertical rough sawn siding, flat roofs, fieldstone and vertical glass window treatment contributes to the country look. The entrance foyer leads straight ahead to the family room that features a fireplace and an all-glass rear wall; to the left is the living room and sunken dining room. On the upper level the sleeping wing, consisting of three bedrooms and two full baths, is angled away from the mainfoyer for privacy and complete quiet. The lower levels include a two car garage, lavatory, recreation room, den and hobby room.

AREA: Living level   1499 sq. ft.
          Sleeping level   1128 sq. ft.
          Lower level   2747 sq. ft.

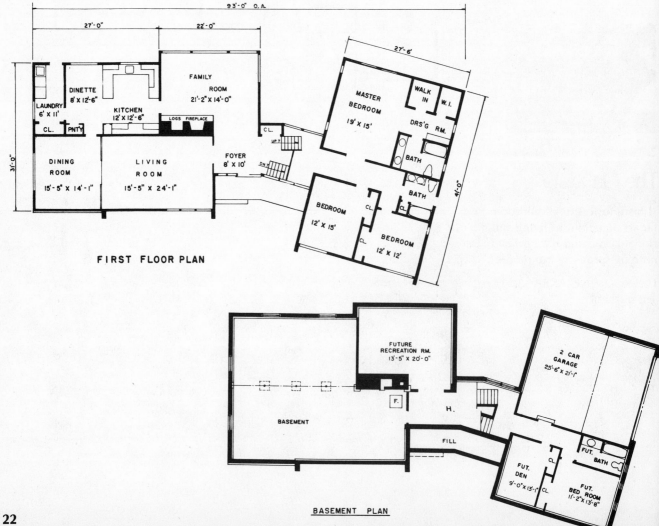

FIRST FLOOR PLAN

BASEMENT PLAN

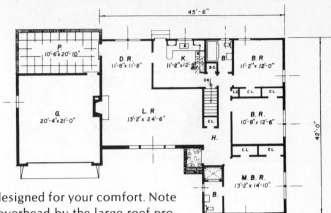

The Larchmont

This spacious six room ranch home has many features expressly designed for your comfort. Note the front entrance protected from the weather on two sides and overhead by the large roof projection. There is a through hall leading directly from the entrance to the kitchen, and the kitchen itself is spacious, with enough room for table and chairs. A large porch off the dining room with access to the garage provides ample sitting area and convenient serving for outdoor dining.

Notice the "L" shaped dining-living room, giving the advantages of the modern open plan, yet providing ample wall space for convenient furniture arrangement.

You will live graciously and well in this home with its private master bedroom, shower bath and enormous double closets. Note that the bedroom lighting is by many high windows, a typical ranch style feature providing ample wall space below for furniture and beds.

TOTAL LIVING AREA: 1,506 sq. ft.

The Oakdale

This unusual three-bedroom two-bath ranch home incorporates the new and increasingly popular arrangement of living and dining rooms in the rear. Combining the best features of the conventional home and the newest in modern living, this plan will prove to you, upon examination, that nothing surpasses our modern day planning for economy and convenience of living.

TOTAL LIVING AREA: 1,515 sq. ft. (excluding porch
& garage)

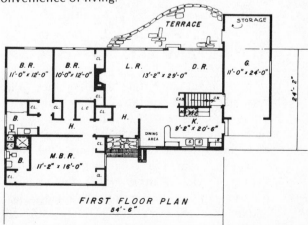

The Barrett

Family comfort counts in this spacious ranch house. Large rooms are featured with each of the three bedrooms "master" sized. The closets are tremendous and there are plenty of them. The economy of back-to-back plumbing affords luxury of a full family bathroom including a vanity powder room alcove as well as a private shower bathroom for the parents' bedroom . A skillfully planned kitchen includes a breakfast alcove which could be curtained to make a full room. The entertaining "L" formed by the living and dining rooms and the family room makes indoor-outdoor living a joy and wonderful windows bring summer breezes and winter sunshine in.

TOTAL LIVING AREA: 1,530 sq. ft. excluding garage.

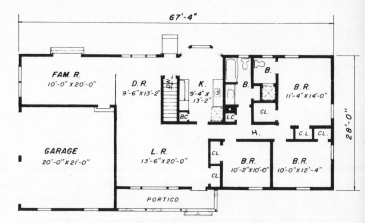

The Ridgewood

This compact ranch style home embodies all the features of its larger sprawling prototype, yet is small enough to fit on the average suburban lot. There are three spacious bedrooms and two well-designed baths. Note the tremendous closet space, particularly the large walk-in type in the master bedroom. The entry-hall closet has plenty of room for family wraps and guest's wraps with no crowding. A very desirable feature is the vestibule entry. Recessed for weather protection, it affords a buffer area for cold winds and wet feet. Note that the kitchen is directly accessible from the hall—which means no traffic through living and dining rooms. There are plenty of work counter areas in this kitchen, lots of cabinets above and below and an alcove corner for breakfasts and lunches. The extra wide living room directly off the entry hall opens through an archway into a spacious dining room which in turn connects the kitchen and porch for convenient serving on summer evenings.

24   TOTAL LIVING AREA: 1,571 sq. ft. (excl. porch & garage)

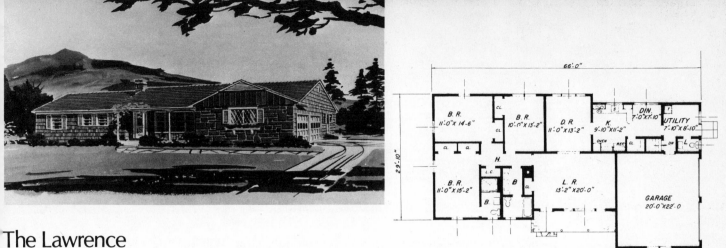

The Lawrence

The warm exterior lines of this ranch plan stand ready to extend a welcome to the family whose home it will become.

All rooms are sized for comfort and arranged for convenience. The utility room at the rear corner serves multi-purpose as laundry room, mud room and foyer area from outside, garage and basement.

There is an abundance of work counter and cabinet space in the kitchen as well as a pantry closet and separate dinette space.

The master bedroom has its private bath and the hall bath is separated into two areas for multi use.

Closet space in all three bedrooms is ample, and in addition, there are a large linen closet in the hall and a guest closet at the entrance.

TOTAL LIVING AREA: 1,540 sq. ft.

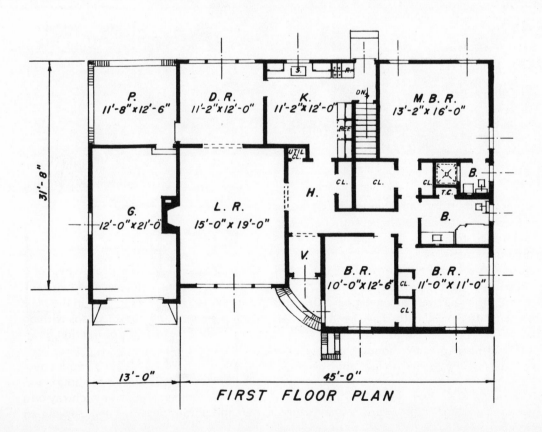

FIRST FLOOR PLAN

25

The Edgecliff

Here is a house, planned for perfection—your dream home come true. Window bright using the new awning windows with large glass areas and full 100% ventilation area, opened by convenient worm gear operators. The master bedroom at front is a parents' dream for space and light—3 way ventilation with all windows set high for privacy and unlimited wall space. All closets are king size, but the large walk-in closet in the master bedroom deserves special notice. There are two full baths, both with vanitories and both conveniently spacious.

TOTAL LIVING AREA: 1,600 sq. ft.

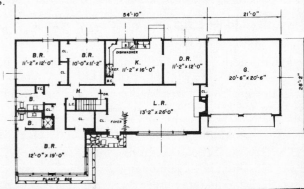

The Edwards

From the picturesque shutters and planting box gracing the front to the breezy rear terrace with its useful barbecue pit for those summer get-togethers, this popular design has answered the dreams of many.

There are three large bedrooms conveniently located near the two full baths and more closet space than you will ever be able to fill. The spacious living-dining room and modern kitchen with its cozy dinette have won wide acclaim.

Completing this wonderful ranch home, the basement level contains a large laundry room with adjoining bath.

TOTAL LIVING AREA: 1,644 sq. ft.

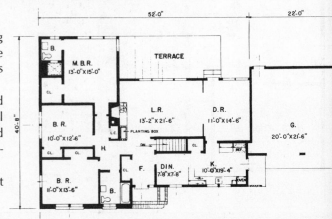

The Dumont

This three bedroom ranch home is designed for informal living, but it has, too, all the regal arrangements of the formal home. A through hall entrance, leading directly to the kitchen area and separating living and sleeping areas, provides all the dignity and grace of well-planned circulation.

The enormous 26' combination laundry-kitchen-dining area is an outstanding feature for informality and convenience. There is plenty of open floor area for the children to play right where mother can keep tab on their activities while doing the washing or preparing meals. Then, too, the kitchen "L" counter can double as a snack bar for those quick pick-me-ups.

The large breezeway off the living room and adjacent to the dinette has ample area for entertaining on warm summer evenings, and it also provides sheltered passage in all weather to the garage. Two full size baths and lots of extra separate closet space for clothes, linen and towels, plus three bedrooms, all with ample room for twin or double size beds and furniture, complete this home—designed for the finest of living in the most pleasing manner.

TOTAL LIVING AREA: 1,618 sq. ft. excl. garage
   & porch.

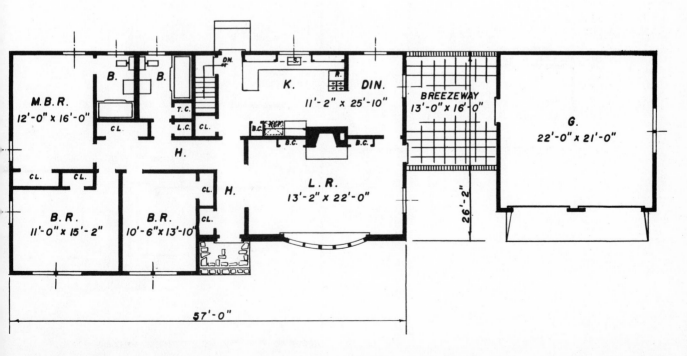

The Marlowe

Here is the complete home you have been looking for, all on one floor.

Entrance to its two car garage may be from any of three sides, depending on your lot size. Upon entering, directly ahead is the living room and dining room with its handsome common window wall and built-in glass china cabinet in the dining room.

From the kitchen with its separate table area, you enter the room of many uses. This family room, opening directly to a covered patio, provides for indoor, outdoor living and entertaining.

In the foyer there is an eye catching brick planter and grilled wall at the living room, which gives this area a very handsome look.

The main bath, with its convenient location, also serves as a guest powder room.

Three spacious bedrooms, including abundant closet space and private bath with stall shower for the master bedroom, will satisfy the family's needs.

TOTAL LIVING AREA: Living area   1,675 sq. ft.
                  Garage   535 sq. ft.

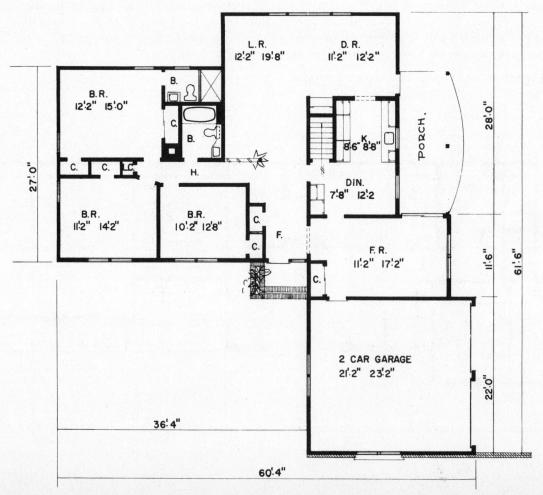

The Stratton

All the comforts of one-level living are found in this rambling ranch house. From the convenient foyer and halls to the spacious garage, here is a house you will be thrilled to call your home.

The peaceful sleeping area, separated from the rest of the house, offers complete privacy for its three leisurely bedrooms and two full baths.

You will be impressed with the magnificent, spacious living area which includes a cheerful living room, a delightful dining room, and a perfectly arranged kitchen.

As an added extra, for those bad weather spells, there is a handy door leading from Dad's den to the family garage.

Numerous closets and windows, and a beautiful porch for those summer days add the final touches to this home designed just for you.

TOTAL LIVING AREA: 1,675 sq. ft.

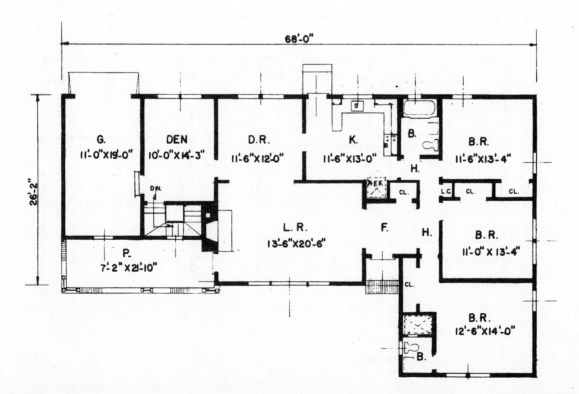

The Wildwood

An enduring quality is presented in the long, low, rangy lines of this ranch home. The entrance is in the shielded corner of the front patio, and inside a smart foyer separates the three-bedroom wing from the living and entertaining areas. And what a living room there is—almost a 25' sweep to the fireplace, and another 25' expanse through the arch of the dining room, from picture window to picture window. Next to the dining room there's an extra, all-purpose room, with two sun-filled window walls. It's convenient from house and garage, and can double as a playroom, study or tremendous television den. Closet walls add spacious storage to the bedrooms, and the family bathroom, economically back-to-back with kitchen plumbing, is supplemented by the master shower-lavatory. The kitchen has a spacious, charming area for dining, as well as cooking in this interesting luxury house.

TOTAL LIVING AREA: 1,710 sq. ft. (excl. garage)

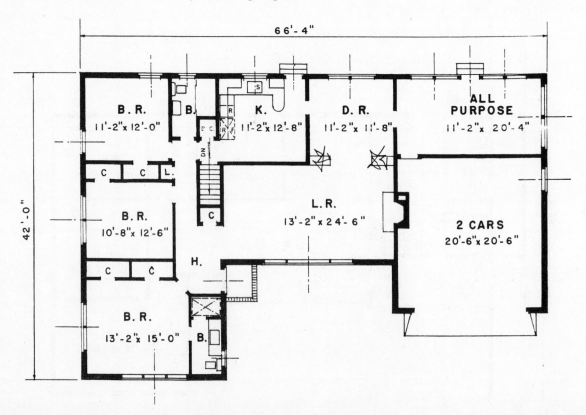

The Whitestone

Being two-faced in this case is not a detriment, quite the contrary as a matter of fact. This "L" shaped ranch home can be set on your lot to present either of two faces to the front, (whichever pleases your individual taste). All rooms are good size and closets are many and large. The living and dining areas combined into an "L" shape make each appear much larger than they really are, and the long kitchen provides ample space for convenient work and breakfast areas. The den is located in such a position that it can serve well as an extra guest room, being directly across the hall from the extra shower-bath-lavatory located next to the kitchen.

TOTAL LIVING AREA: 1,715 sq. ft.

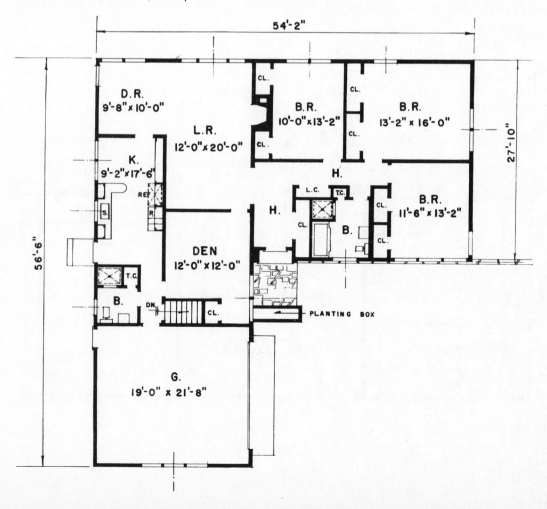

The Hampton

If you have been thinking, at one time or another, of living in a spacious and convenient ranch home, here is a traditional three-bedroom design that has the ingredients of a much larger home and will fit perfectly into almost any setting.

In this version, the covered entrance provides a welcome introduction to the house at the center of the plan, from which you can reach any room with a minimum of steps from the adequate foyer that serves as a traffic control center.

TOTAL LIVING AREA: 1,720 sq. ft.
    Basement   1,630 sq. ft.
    Garage   484 sq. ft.

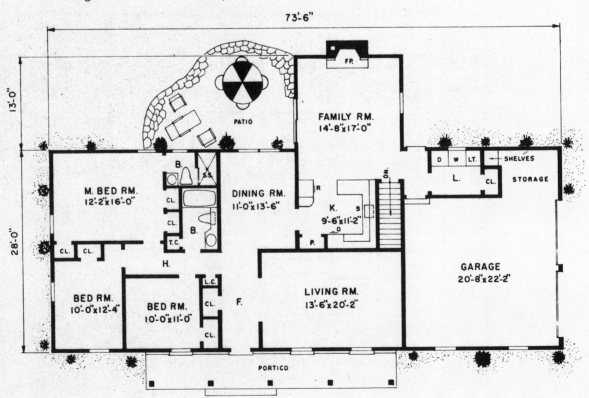

The Nantucket

This long, low, rambling dwelling appears to be a conventional ranch, but has two extra bedrooms and a bath upstairs . . . living room has a large picture window set in a box bay . . . beyond the living room is the dining room, entered through an arched opening . . . family room, kitchen, laundry room are in line in the back part of the house . . . corner fireplace in the living room is visible from the foyer . . . a hall leads to three bedrooms and two baths on the first floor. The second floor can be added at a later date.

AREA: 1st floor   1772 sq. ft.
         2nd floor   483 sq. ft.

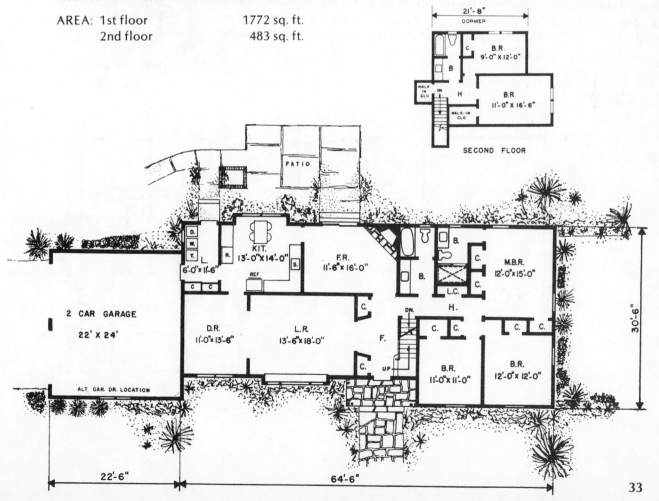

The Frontenac

Luxurious living in the contemporary manner is provided in the ultimate in this distinctive one story, three bedroom "rambling" ranch design. Despite its somewhat expansive appearance, it actually is a compact plan with 1,814 square feet of area with eye catching features of the exterior such as vertical stone piers between the living room picture plate glass windows, natural rich color red-wood vertical siding, stone veneer, low hipped roof and the stone planter at the covered entrance portico.

With simplicity as its theme, it fulfills the aim of producing a design with plenty of exterior sparkle.

TOTAL LIVING AREA: 1,814 sq. ft.
    Basement   888 sq. ft.
    Garage   529 sq. ft.

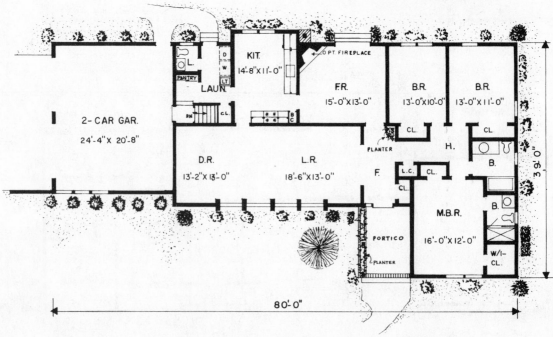

The Bayshore

This hip roof ranch has an exterior that tastefully blends walls of brick and wood shingles.

The entrance foyer is centrally located between the living-dining areas and the sleeping quarters.

The family room, kitchen and laundry room are in line in the back part of the house. Fully equipped with modern appliances and plenty of counter and storage space, the kitchen has a dinette area which looks out to the rear garden and patio. To the left of the kitchen is the laundry room with pantry and adjacent lavatory. There are doors in the laundry leading to the rear yard and to the two-car garage.

To the right of the foyer is a hall leading to three bedrooms and two baths. The two front bedrooms will accommodate most furniture arrangements. The master bedroom, also with excellent wall space, has two closets and its own bath, the latter tiled, with its basin set in an attractive, built-in vanitory, and with a stall shower.

This ranch house is sure to be a leader in any community.

TOTAL LIVING AREA: 1,830 sq. ft.

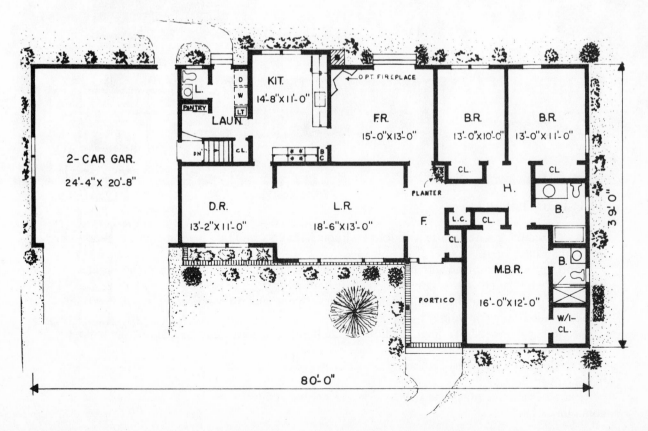

The Riverdale

Here is a ranch house that specializes in living conveniences. Some of the many features which afford these conveniences are the walk-in closets with plenty of space—no crowding of clothes ever—and a long sliding-door linen closet with many shelves providing all the storage space you will ever need. There is a separate vanity in the main bath, a large vanitory in the master bath and three very large bedrooms. Notice the small alcove off the kitchen leading from outside directly down to the basement. The living and dining rooms form a spacious bright and airy "ell" with views front and rear through large window areas. As a final touch there is an ample sized two car garage attached. It is accessible through the long open porch area directly off the dining room.

TOTAL LIVING AREA: 1,860 sq. ft.

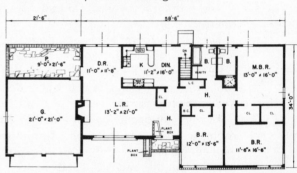

The Farmstead

Here is a rambling ranch house with four bedrooms, laundry and family rooms plus 2½ baths, all in less than 1,900 sq. ft.

Wood shingles, combined with board and batten exterior siding along with a graceful entrance portico give this home an eye appeal that will be hard to beat.

Adjoining the kitchen, the paneled family room with colonial fireplace, and glass sliding doors, allows for ideal "outside-inside" living.

The kitchen with separate dinette, adjoining the laundry, lavatory and two car garage make for ideal traffic circulation.

All bedrooms have oversized closets, excellent wall space and more than adequate natural light and ventilation.

After studying this plan, one will have to agree that its design is sure to be a leader in any community.

TOTAL LIVING AREA: 1,898 sq. ft.

The Curran

Modern living in the atmosphere of the old colonial is found in this popular design.

Four spacious bedrooms with ample closet space are conveniently located to bring you the seclusion and privacy you desire.

A built-in two-car garage is found in the basement.

TOTAL LIVING AREA: 1,885 sq. ft.

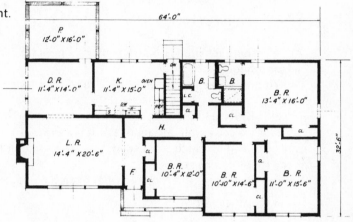

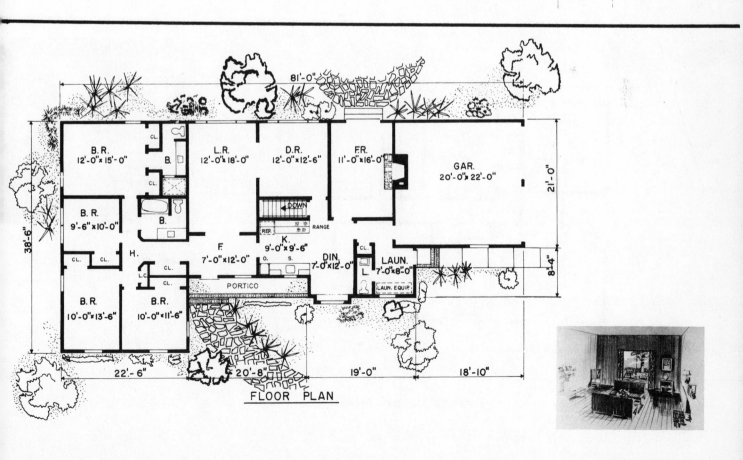

FLOOR PLAN

The Lewiston

A refreshing exterior is boasted by this popular ranch design which will make it a hit in any neighborhood.

Including three airy bedrooms, a modern kitchen and the combination sunken living-dining room.

Gable roofs and shingles highlight the exterior of this roomy ranch home.

Inside we see a well designed layout featuring large rooms each conveniently located. A library is also provided which, if needed, may serve as another bedroom. Two full baths back-to-back make for plumbing economy.

These and many other features make this a popular home.

TOTAL LIVING AREA: 1,900 sq. ft.

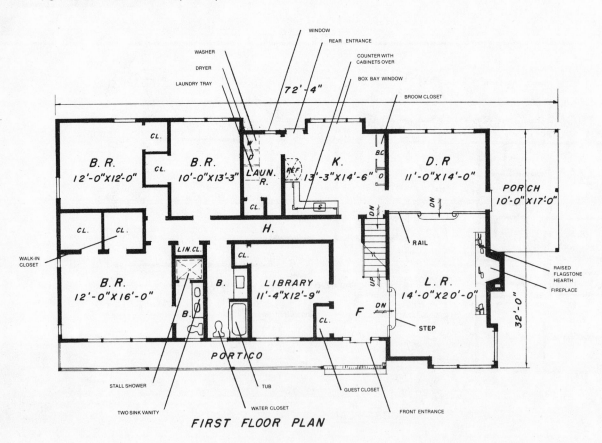

FIRST FLOOR PLAN

The Lexington

The flavor of old New England is present in the lines of this ranch home for today's living.

Notice that the kitchen is in the front of the house. This location provides many advantages both economical and convenient. All plumbing is at the front of the house, meaning shorter runs to street connections. Circulation within the house is ideal, for the kitchen is directly adjacent to the dining room, porch, exterior and entrance foyer; connecting to all these areas without using other rooms as passageways.

Dining room and living room in the rear provides privacy and a controlled view of your own property area.

Three bedrooms and two baths complete the living area of this home, but there is more—in addition to a full 2 car garage located in the basement, we also find a wonderful recreation room, laundry area and a lavatory, in addition to tremendous area for the heating unit and much storage and work shop area.

An alternate arrangement on the blueprints shows the garage at the front of the house and the recreation room at the rear.

AREA: First floor 1,793 sq. ft. (excl. porch)

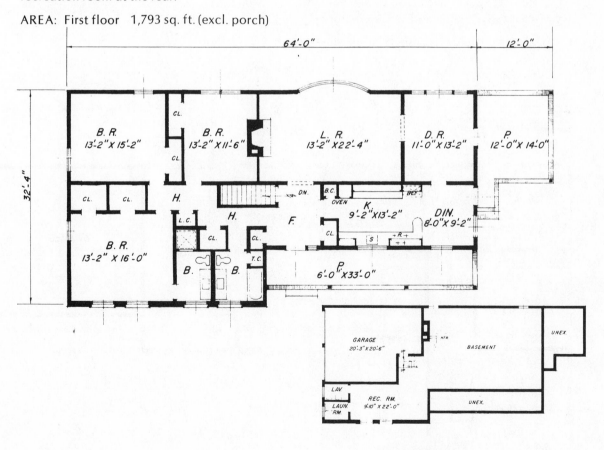

BASEMENT & GARAGE PLAN

The Devons

Three large bedrooms and two full baths, one with a stall shower, line up to form a discrete and serene sleeping area.

Surveying the rest of the floor arrangement, we find a relaxing living room, the family room, and a complete step-saving kitchen adjacent to the sunny dining room.

Spaciousness keynotes each and every room.

TOTAL LIVING AREA: 1,925 sq. ft. (excluding garage)

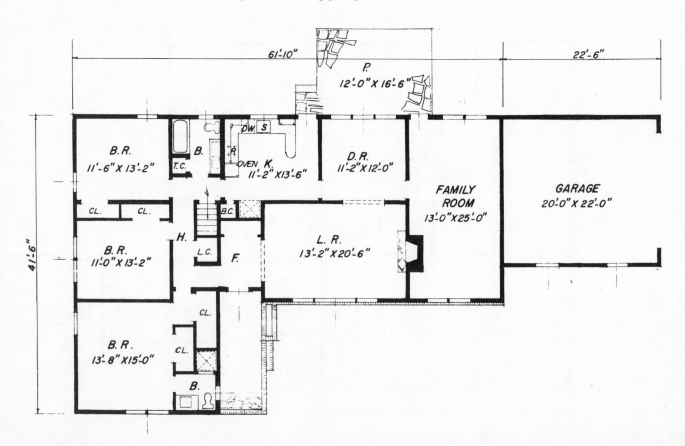

The Dorian

This three bedroom ranch home is designed for informal living, but it has, too, all the regal arrangements of the formal home. A through hall entrance, leading directly to the kitchen area and separating living and sleeping areas, provides all the dignity and grace of well-planned circulation. Two full size baths and lots of extra separate closet space for clothes, linen and towels, plus three bedrooms, all with ample room for twin or double size beds and furniture, complete this home—designed for the finest living in the most pleasing manner.

TOTAL LIVING AREA: 1,945 sq. ft.

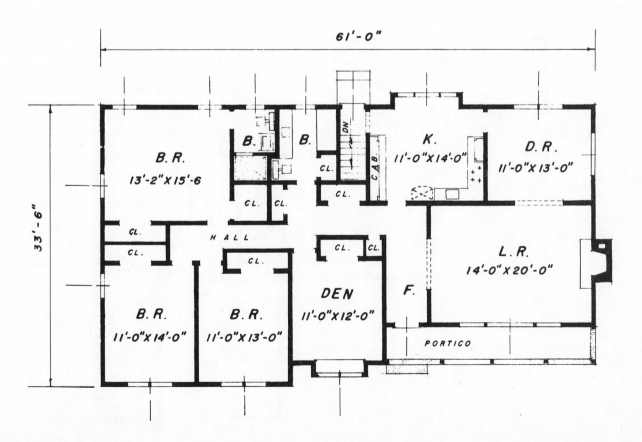

The Dunbar

Condensed in this compact ranch home, formed in the popular "L" shape, are all the features found in many larger and more expensive homes. It has tremendous closet space, conveniently located throughout the entire house, three ample sized bedrooms, with a spacious family room off the kitchen. Leaded glass windows give this home the typical spreading look of a colonial.

TOTAL LIVING AREA: 1,945 sq. ft.

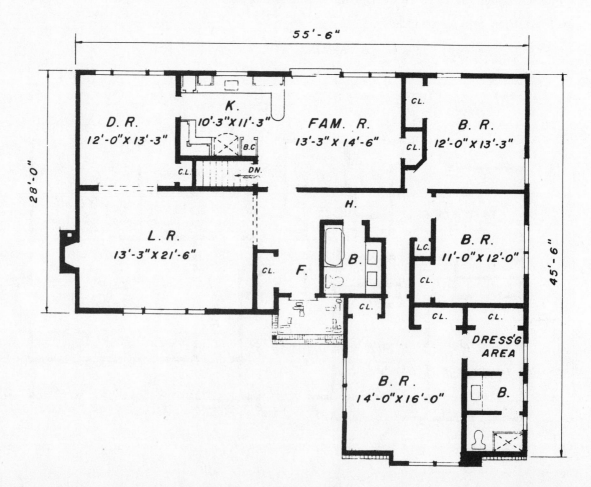

The Pearson

This homey design will form a perfect picture of beauty in any neighborhood.

The basement contains a roomy two car garage and ample room for a future recreation room.

A convenient lavatory is found off the kitchen and foyer. The dining area is highlighted by a picturesque box bay.

TOTAL LIVING AREA: 1,950 sq. ft.

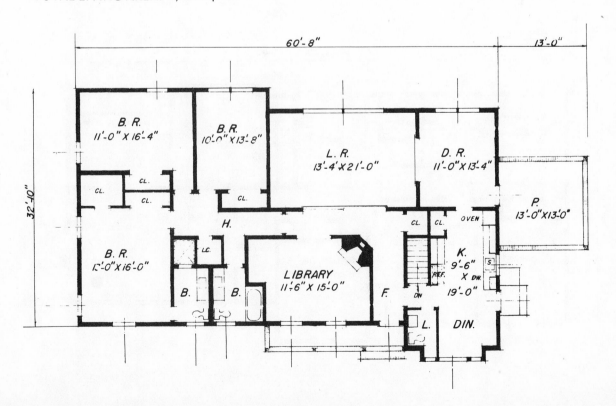

The Lamont

Four large bedrooms, two full baths and many convenient closets form the discrete and serene sleeping area.

Convenience and spaciousness keynote the rest of the house. Living room with fireplace, a large cheery kitchen with dining alcove and dining room access to patio enjoyment complete the design for gracious one level living.

TOTAL LIVING AREA: 1,955 sq. ft.

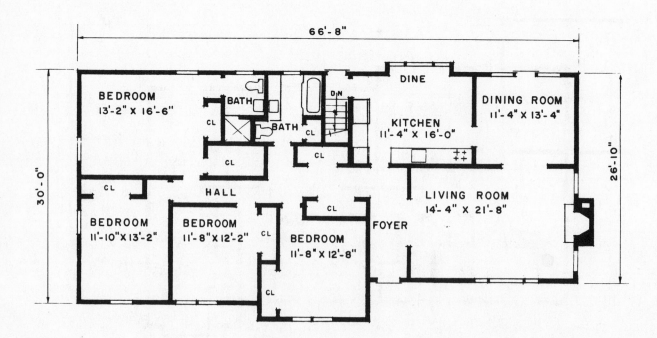

The Clarke

This contemporary styled ranch plan offers many desirable features for today's living.

The large utilities room on the living level with convenient lavatory provides direct access to kitchen, outdoors and basement.

The open "L" shape of living-dining rooms and the spaciousness of the entrance foyer accent the free feeling of modern uncluttered living.

TOTAL LIVING AREA: 1,985 sq. ft.
(excluding porch & garage)

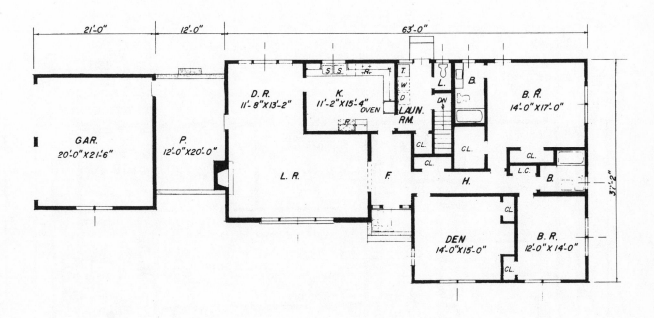

The Chathampton

This is a well proportioned three bedroom, many closeted, two bath ranch home with the additional feature of a den off the living room. This bonus room adds to the overall length of the house and really emphasizes the "spread out" appearance so typical of ranch homes. A porch set in the "ell" formed by dining room and den fills out the proportions of this home for regal ranch type living.

TOTAL LIVING AREA: 2,029 sq. ft.

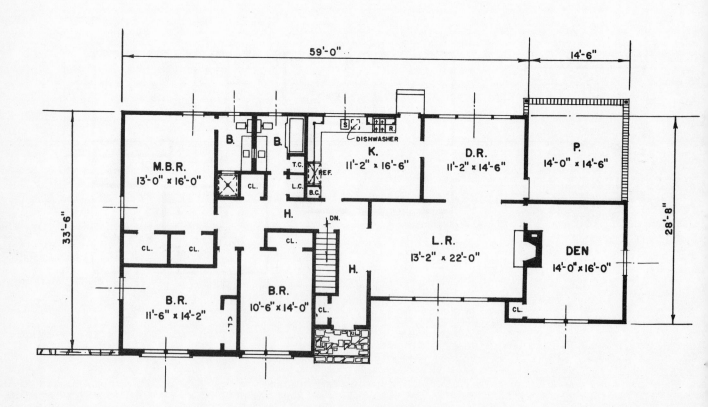

The Villanova

This latest design with that increasingly popular Spanish-Colonial influence makes you feel like a nobleman of a past era with all the advantages of modern materials and conveniences.

Accentuated by the low-walled wrought-iron and main entrance gates, rough stucco finish, projecting stained wood beams, arched windows, paved courtyard with circular fountain, ceramic tile entrance foyer floor and a screened-in patio large enough to include the optional swimming pool lend an exotic air of a Spanish villa to this design.

Although the plan is basementless, a full or partial basement is possible with the basement stair located where the large utility closet is shown in the laundry area. The laundry is complete with washer, dryer and accessible to the two car garage.

This design, pure or modified, seems perfectly adapted to today's living and has a romance that is typical of the traditionally Spanish-styled homes.

AREA: Living area   2,060 sq. ft.
        Garage   460 sq. ft.
        Screened patio   870 sq. ft.

POOL

PATIO

M. BED RM.
12'-4"x16'-0"

DRESS'G AREA

B.

WALK-IN CLOSET

S.S.

CL. CL.

BED RM.
9'-4"x12'-0"

B.

H.

L. CL.

BED RM.
9'-4"x12'-0"

CL.

CL.

DINING RM.
11'-0"x12'-8"

K.
9'-8"x11'-0"

CL.

UTILITY OR OPT. STAIRS

L.T.

W.

D.

L.

CL.

DEN
10'-0"x14'-0"

FAMILY RM.
13'-0"x23'-0"

FIREPLACE

LAV.

CL.

LIVING RM.
13'-0"x17'-8"

F.

PATIO

GARAGE
21'-2"x22'-0"

FOUNTAIN

COURT

67'-2"

57'-2"

FLOOR PLAN

The Norwood

A ranch house consisting of seven rooms and including a two-car garage and two baths will be your family's favorite.

Three of the rooms are bedrooms. The den might serve as an occasional fourth bedroom when there are extra guests. Ordinarily it is more apt to be a television room and is shaped accordingly — 24½' long. The other rooms include an efficiency U-type kitchen, dining room and an enormous living room 29' long. There are no less than ten closets, and book cases flank the living room fireplace. In the same rooms two virtually unbroken walls permit attractive placement of large pieces of furniture.

TOTAL LIVING AREA: 2,087 sq. ft. (excluding garage)

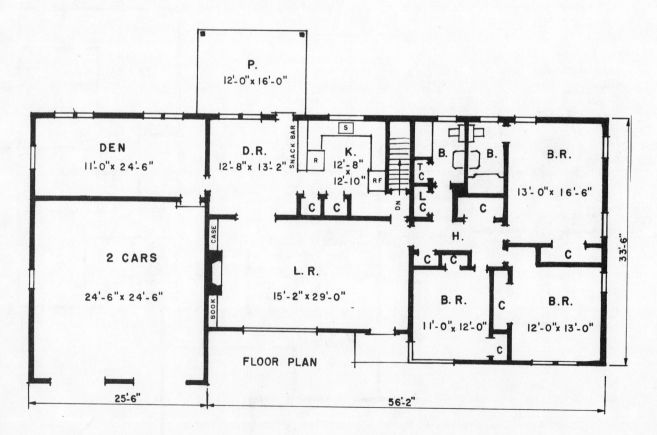

P. 12'-0"x 16'-0"

DEN 11'-0"x 24'-6"

D.R. 12'-8"x 13'-2"

K. 12'-8" x 12'-10"

B. B.

B.R. 13'-0"x 16'-6"

2 CARS 24'-6"x 24'-6"

L.R. 15'-2"x 29'-0"

H.

B.R. 11'-0"x 12'-0"

B.R. 12'-0"x 13'-0"

FLOOR PLAN

25'-6"   56'-2"   33'-6"

The Baxter

Beauty and simplicity, together with thoughts of economy, have been united to create this three bedroom stone veneer and wood shingle design. The unusual cozy layout has all the privacy of a two story house with the sleeping area separated from the living activity.

The large living room and dining room give a spaciousness usually found only in much larger and expensive homes. A two car garage is conveniently located under the bedroom area.

Dress up the planting box at the entrance with your favorite flowers, and this house will stand out among others with your own personal touch.

TOTAL LIVING AREA: 2,125 sq. ft.

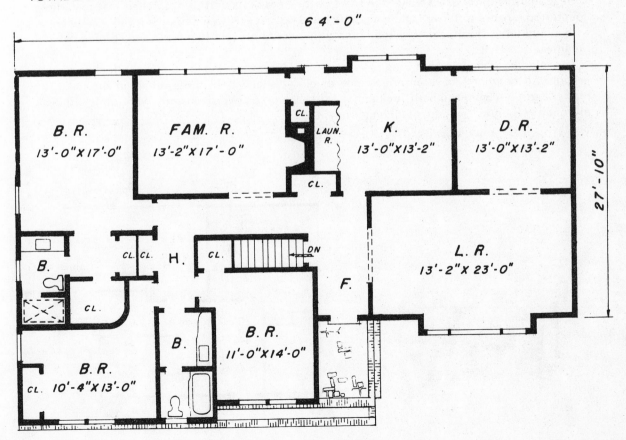

The Sturbridge

Entering beneath the covered portico into a larger foyer, one moves straight ahead to the living room, to the left and the four-bedroom wing, or to the right and the kitchen. This distributes traffic effectively.

The living room has plenty of wall areas suited for varieties of furniture arrangements, but the feature is the fireplace wall. It's made of floor-to-ceiling brick throughout its length, with an oversized, extended flagstone hearth. The back of the fireplace gives the dining room an interesting bricked projection. Next to the dining room is a wood-paneled family room, with full-height sliding glass doors leading to a rear patio.

Headquarters for housekeeping is in front of the home in an efficient kitchen that includes a countertop range, a wall oven and a separate breakfast area. Also in this area is the service entry, with a mud-room closet, lavatory and laundry that contains a washer, dryer and storage cabinet facilities.

The four-bedroom, two-bath wing has a hall flared out at the end to eliminate a feeling of congestion. All bedrooms have an abundance of closet space, with three others in the hall. The master bedroom has a full bath, including a stall shower, an oversize vanity with full-length wall mirror and separated, water-closet compartment.

TOTAL LIVING AREA: First floor   2,134 sq. ft.

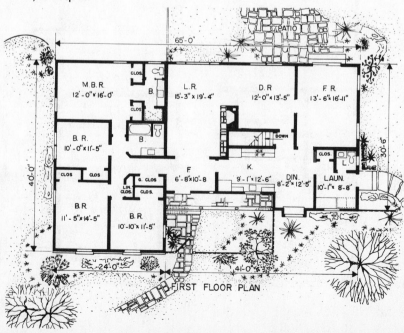

FIRST FLOOR PLAN

The Blaine

Here is an exclusive ranch design featuring a circular foyer. A handy powder room and a large living room are only a few of the features to be found in this contemporary design.

The bedroom area contains three large bedrooms, two baths, and ample closet space. Here we also find a cozy den for Dad. A large kitchen and dining room open out to the terrace with its handy barbecue pit.

TOTAL LIVING AREA: 2,145 sq. ft.

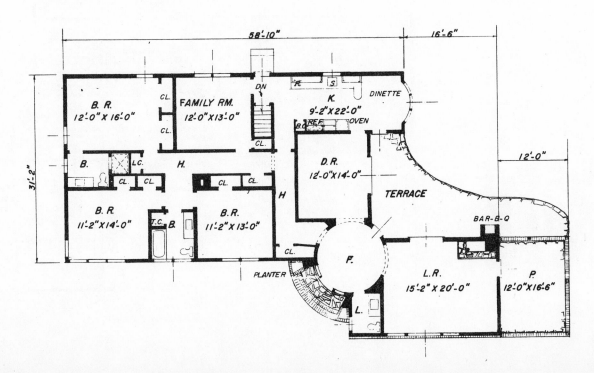

The Hacienda

While the exterior architectural details of this three bedroom one-story house adhere to the Spanish motif, the floor plan is arranged for present-day living.

The entrance is most impressive with wrought iron gates leading to the front door through a private, typical "ranchero" treatment.

This house is for the family that wants the conveniences of today wrapped in a look of yesterday.

TOTAL LIVING AREA: 2,145 sq. ft.

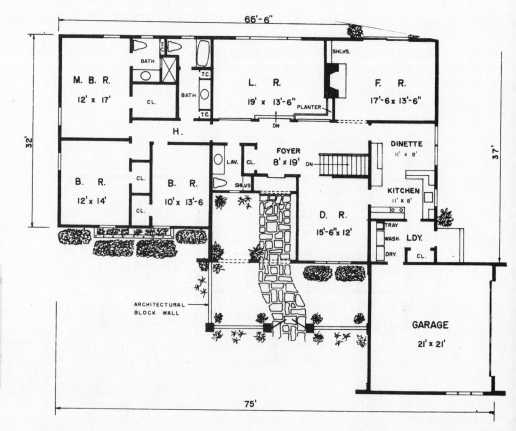

The Regent

Gable roofs, stone veneer and a modern retaining wall form a beautiful exterior to accent this captivating ranch design.

The floor plan reveals back-to-back plumbing, a cozy fireplace in the den, and numerous closets and sunny windows.

A built in two car garage is to be found in the basement layout along with ample space for a future recreation room.

TOTAL LIVING AREA: 2,145 sq. ft.

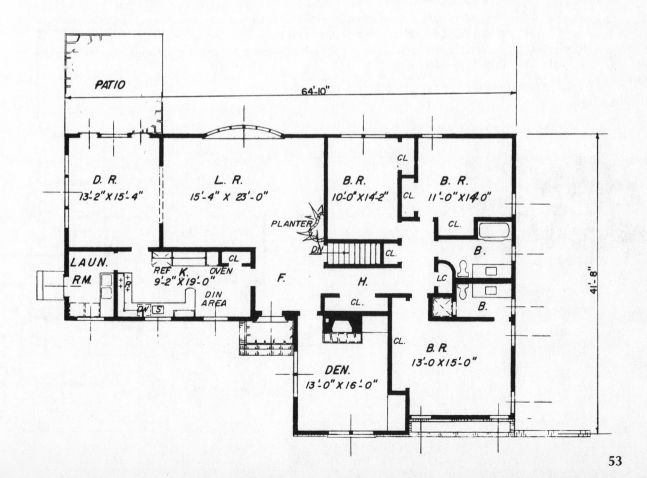

The Menlo

Restraint and sophistication are combined in this contemporary design which utilizes vertical siding . . . special windows and interesting roof to produce a house that is different, yet has clean architectural lines. Among the eye-catching features of the exterior of this eight room, three or four bedroom ranch design is a trellised roof which casts interesting shadows over the double-door entrance . . . and the tapered skylight roof which admits natural light during the day into the central foyer of the bedroom wing.

TOTAL LIVING AREA: 2,218 sq. ft. (excluding garage)

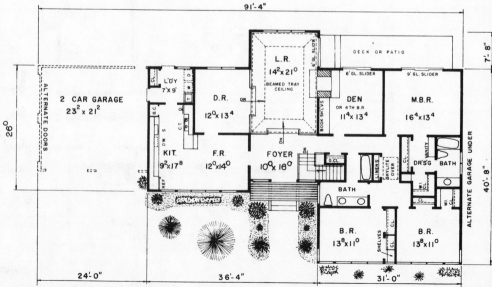

The Rutledge

This "L" shaped ranch design provides comfortable living for all. Additional living area has been provided with the inclusion of a family room and T.V. room.

Many features are included in this design to bring you a home that you can truly be happy to own.

TOTAL LIVING AREA: 2,290 sq. ft. (excl. garage)

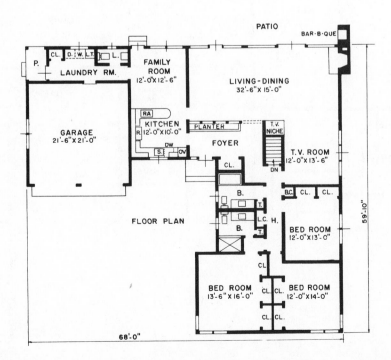

The Vancouver

Reminiscent of French chateau architecture, this one and one-half story, traditional home has the quiet dignity of country living. Entrance to this lovely home is through a garden court, and the double door entry leads straight ahead to the spacious living-room. The kitchen-dinette is designed for every modern built-in convenience, and its size makes meal preparation much easier; adjacent to it is the wood-paneled family room that features a raised-hearth fireplace and sliding glass doors to the rear patio. The three bedrooms are of modest size, but the master has its own private stall shower bath, and one of the front bedrooms is fortunate to have a view of the front courtyard. Two additional bedrooms and bath can be built on the second floor, at a future date, if so desired.

TOTAL LIVING AREA: 2,300 sq. ft.
    Second floor   493 sq. ft.
    Garage   500 sq. ft.

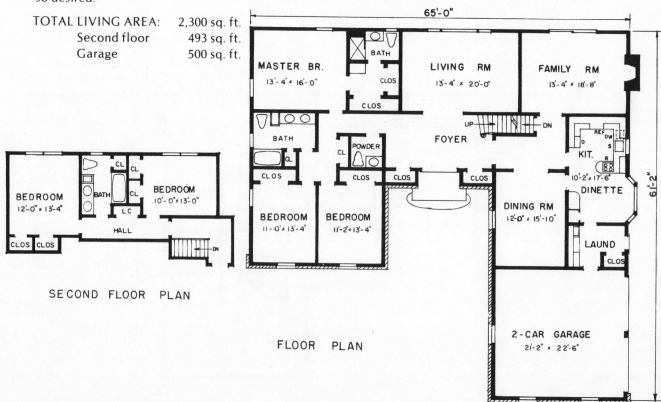

SECOND FLOOR PLAN

FLOOR PLAN

The Gladstone

The quaint California character of this dwelling makes it an asset to any section of homes.

All the features of the "rambling ranch" are incorporated herein—from its sunken living room with sliding glass wall for indoor-outdoor living to its definitely separated living and sleeping areas.

Picture a wide sweeping driveway leading to the impressive entrance porch.

Colorful plants border your steps as you walk past the picturesque bay window and come upon the dainty detail of the doorway set back in its own secluded niche.

TOTAL LIVING AREA: 2,310 sq. ft. (excl. porch & garage)

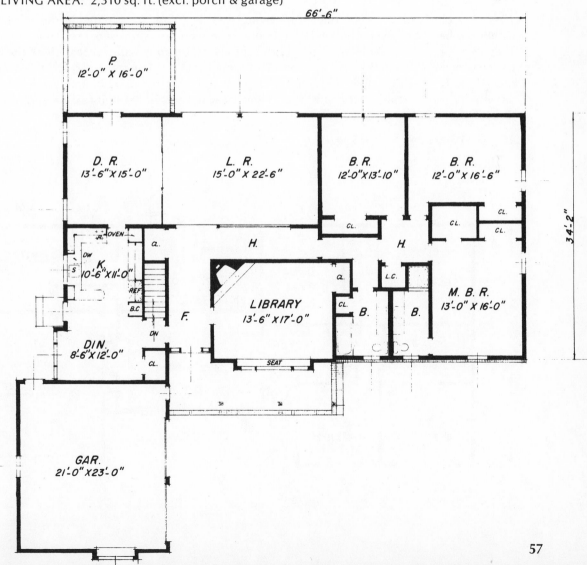

The Fairview

Here is a plan with the new concept for living. Closets galore are situated at the entrance and within the bedroom wing. The kitchen is a dream in work area and separate breakfast alcove. There is a lavatory conveniently located just off the kitchen and readily accessible to the rear entrance which is through the activity room. This room is down a few steps and sports a bright and rugged tile floor. This is an ideal room for children's play or party and certainly will be used by parents for less formal entertaining. The roomy 2 car garage is a built-in feature and has access through the house for rainy weather.

TOTAL LIVING AREA: 2,375 sq. ft.

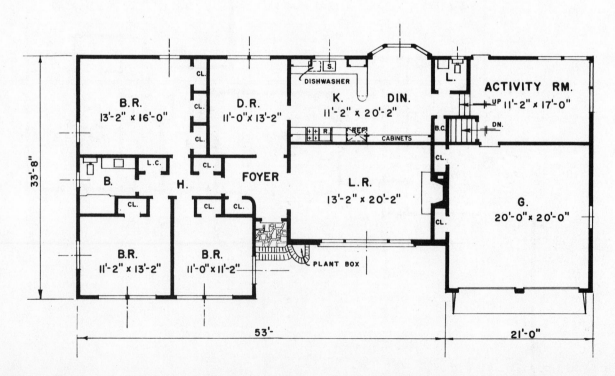

The Covington

Traditional in appearance, with an air of warmth and comfort, this ranch would be an attractive addition to any community.

A covered, long portico leads to the centrally located foyer. Directly to the rear of the foyer is a sunken living room and adjoining dining room with a wrought iron rail acting as a divider.

To the right of the foyer the wood-paneled family room has a brick-faced fireplace that is visible almost immediately after entering the front door.

Between the combined kitchen-dinette and the garage is the laundry room which contains a closet for cleaning supplies and equipment. The laundry room has two doors and one to the garage.

The two rear bedrooms, with the main bathroom between, will accommodate most furniture arrangements. The master bedroom with its dressing area and spacious closets, one of which is a walk-in type, complete this lovely ranch.

TOTAL LIVING AREA: 2,336 sq. ft.

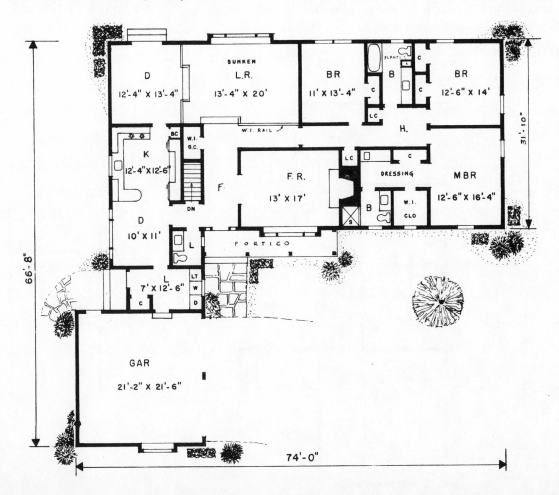

The Sun Valley

Picture this dream house on your lot, a long, flower-edged, curving driveway up the front to the main entrance. A well recessed and weather protected entry leads into a separate entrance foyer. Access is provided from this foyer to the large recreation room on the lower level and garage. Only three steps up from the foyer leads you into a spacious living room and dining room both with a beamed cathedral ceiling. The kitchen and family room are combined. There are 3 bedrooms also on this level and room for two additonal bedrooms on the upper level for future expansion.

TOTAL LIVING AREA: 2,595 sq. ft.

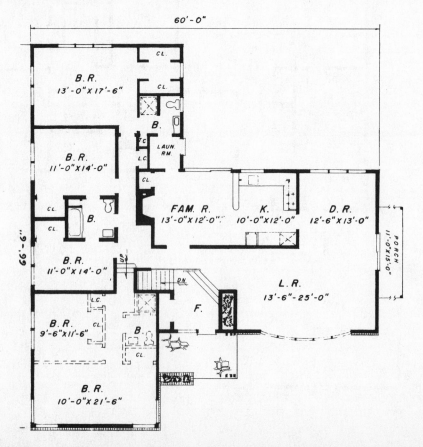

The Lambert

Interesting angles and clerestory windows, softened by the use of fieldstone veneer and redwood vertical siding, give this gentle three bedroom contemporary ranch a distinctive modern look with all the warmth of traditional styling.

Of special interest is the wood-paneled family room that features a stone-face circular corner fireplace with a raised hearth and a built-in log storage bin.

This inviting and distinctive design affords an opportunity for a life style of warmth, flexibility and comfort.

TOTAL LIVING AREA: 2,660 sq. ft.
    Basement   1,700 sq. ft.
    Garage   530 sq. ft.

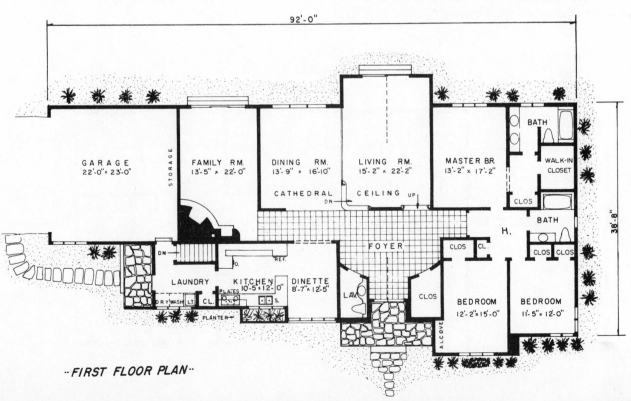

·· FIRST FLOOR PLAN ··

The Hallmark

The comfort and convenience of the ranch style home are unsurpassed. Here is a plan combining all the attributes of up-to-date planning for modern living, yet presenting the formal character of tradition in its outward appearance.

The new trend for outdoor living is accented here by a large sliding glass wall in the den for unrestricted flow between indoors and outdoors.

The master bedroom suite, consisting of dressing alcove, closets and private shower bath, is accompanied by two additional master sized bedrooms and a second full bath, all well supplied with ample closets. A separate lavatory is located convenient to the kitchen, den and living room. The kitchen is arranged for step saving meal preparation, and the breakfast area is located in a bright windowed corner.

A maid's room and bath, laundry room at the service entrance and a spacious 2 car garage finally complete this wonderfully liveable home.

TOTAL LIVING AREA: 2,708 sq. ft.

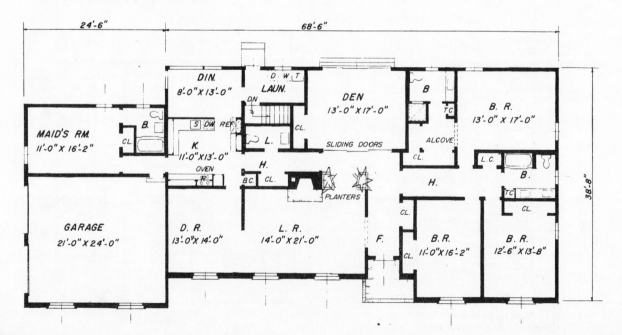

The Aspen

Modern and up-to-date is the best way to describe this spacious contemporary four bedroom ranch that is highlighted by the dramatic split-roof lines, fieldstone veneer, vertical redwood siding, clerestory and vertical window treatment. The double door entry gives a spacious feeling the moment you enter.

The living area includes a large sunken living room, dining room, kitchen-dinette, laundry and a wood paneled family room that features sliding-glass doors to the sun-deck and a see-through fireplace.

Designed to contribute to a feeling of personal luxury, the master bedroom suite has access to the outdoor wood deck, has two closets, one a walk-in, and a complete stall shower bath. The other three bedrooms are served by the main bath, which has a stall-shower, tub and a full length mirrored double-basin vanity.

AREA: First Floor   3193 sq. ft.
        Basement   1943 sq. ft.
        Garage   1250 sq. ft.

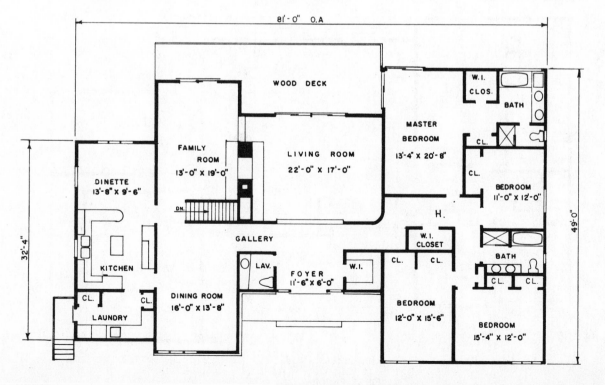

The Escondido

Spanish architecture has maintained a high level popularity through the years, and it is easy to see why. The use of varied materials, design elements and detailing always create a distinctive exterior. This one story "ranchero" is no exception. The stucco finish, arched casement windows, balconied windows and the double carved-paneled entrance doorway under the weather protected walkway are fully compatible with the requirements of contemporary American living.

The "cathedral-ceiling" living room spans 35 feet from the front to the rear patio, which is accessible by a pair of sliding glass doors. To the right is the kitchen-dinette that features an island counter, the laundry and an oversized two car garage with ample storage facilities.

The three bedroom sleeping area consists of a luxurious master suite with a room-size walk-in closet, three additional closets, sunken Roman bathtub, double basin vanity and tiled shower stall. The other two bedrooms are conveniently located to the compartmentalized main bath.

TOTAL LIVING AREA: 3,260 sq. ft.
Garage 680 sq. ft.

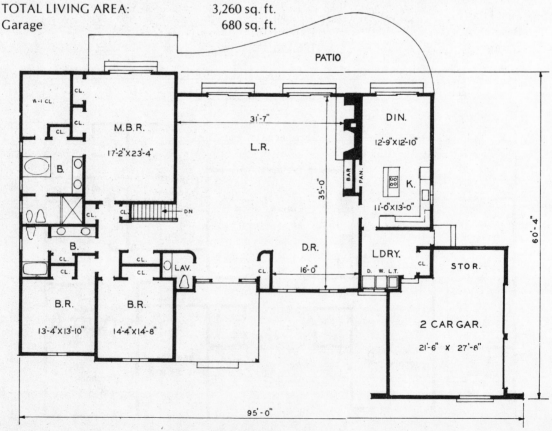

FIRST FLOOR PLAN

EXPANDABLE RANCH

The half-a-story usually refers to an attic which can be finished at the time of the original construction or later on. Often has master bedroom on first floor, children's bedrooms upstairs. Provides extra storage space under eaves. Has knee walls and sloping ceilings upstairs. Most one and one-half story houses have traditional details in the Cape Cod style.

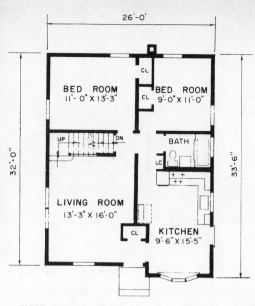

FIRST FLOOR PLAN

SECOND FLOOR PLAN

The Dennis

"Deceiving looking" from the exterior, within lie four large bedrooms with ample closets.

Two full baths are conveniently located outside the bedroom doors.

There is an abundant work area in the kitchen with its connecting dining area which seems to say roomy.

All rooms are sized for family living including the spacious living room.

The needs of comfort, convenience and practical living are combined in this home designed for a narrow lot.

AREA: First floor   832 sq. ft.
        Second floor   448 sq. ft.

The Williamsburg

Traditional beauty and comfort in the Cape Cod style are set forth in this homey five room house with its expansion attic. A wonderful breezeway and garage expand this compact arrangement to provide the luxury look so typical of the New England Architecture.

AREA: First floor   870 sq. ft.
        Second floor   473 sq. ft.

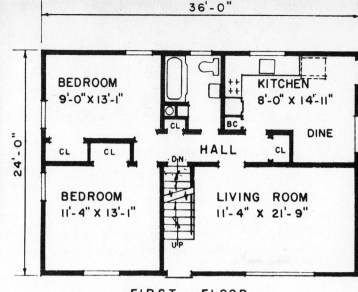

FIRST FLOOR

BEDROOM
9'-0" X 13'-1"

KITCHEN
8'-0" X 14'-11"

CL

BC

CL

DINE

HALL

CL

BEDROOM
11'-4" X 13'-1"

DN

LIVING ROOM
11'-4" X 21'-9"

UP

36'-0"

24'-0"

The De Witt

This compact charmer has everything planned to make a family feel right at home and yet is designed to grow with you and your family. The convenience of the covered entrance and shaded living room windows add to the charm of the contrasting exterior.

The kitchen is unusually large and efficient with ample working area plus a separate dining area. Sharing a closet wall, the two bedrooms are cross ventilated for year round conditioning. As your family grows you'll appreciate the charming dormer bedrooms and second bath.

AREA: First floor   864 sq. ft.
         Second floor   402 sq. ft.

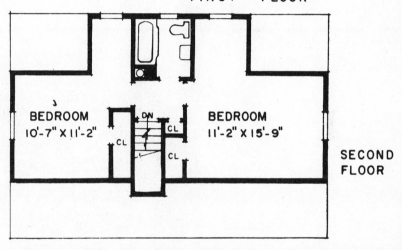

BEDROOM
10'-7" X 11'-2"

DN

CL

CL

CL

BEDROOM
11'-2" X 15'-9"

SECOND FLOOR

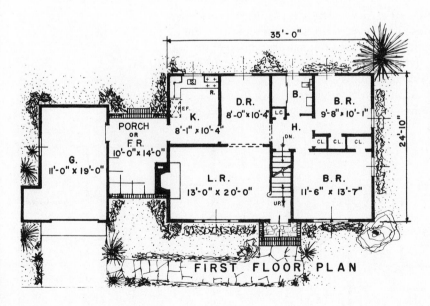

35'-0"

24'-0"

D. R.
8'-0" X 10'-4"

B.

B. R.
9'-8" X 10'-1"

K.
8'-1" X 10'-4"

REF.

LC

H.

PORCH
OR
F. R.
10'-0" X 14'-0"

DN

CL CL CL

G.
11'-0" X 19'-0"

L. R.
13'-0" X 20'-0"

B. R.
11'-6" X 13'-7"

UP

FIRST FLOOR PLAN

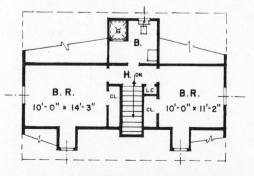

B.

B. R.
10'-0" X 14'-3"

H. DN.

CL

LC

CL

B. R.
10'-0" X 11'-2"

SECOND FLOOR PLAN

The Quaker

A lot as small as 50' could easily contain this home without the garage and porch. The porch could be added later at the rear of the house if your budget allows. Future expansion for two additional bedrooms and a bath is also practical upstairs in this compact home. Basically the four and a half rooms and bath form a wonderfully compact plan with minimum of hall space and a convenience of liveability not often found in larger homes. Notice the luxury features not usually common to small homes such as a full size fireplace, window wall in the living room and a full size bath (no tub-under-the-window arrangement). For a small family with expectations of growing, this is financially an economical initial step with unlimited possibilities for future expansion.

AREA: First floor   877 sq. ft.
       Second floor   450 sq. ft.

The Tennyson

Truly a dream cottage, with its fireplace walls pierced by diamond-paned windows turned to the street, this expansion house is as practical as a house can be. There's fireplace charm in the living room as well as outside, but this spacious room, with the squared dining room opening into it, offers plenty of space for easy hospitality. The kitchen's a step-saver, compact and efficient. Sharing a convenient bath are the two bedrooms, with three windows in one, and cross ventilation for both. Upstairs two more bedrooms are planned, each with lovely dormer alcoves. The second bathroom is over the kitchen plumbing for economy, and there's another linen closet upstairs to save every extra step.

AREA: First floor   990 sq. ft.
       Second floor   475 sq. ft.

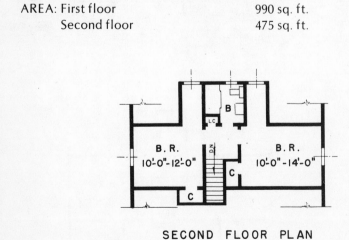

SECOND FLOOR PLAN

FIRST FLOOR PLAN

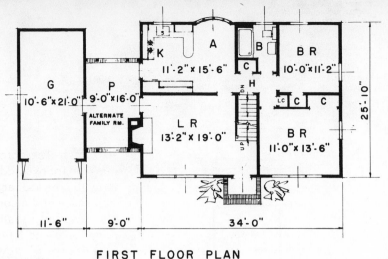

K
11'-2" x 15'-6"

A

B

B R
10'-0"x11'-2"

G
10'-6" x 21'-0"

P
9'-0"x16'-0"

ALTERNATE
FAMILY RM.

H

LC

C

C

L R
13'-2" x 19'-0"

UP

B R
11'-0"x 13'-6"

25'-10"

11'-6"   9'-0"   34'-0"

FIRST FLOOR PLAN

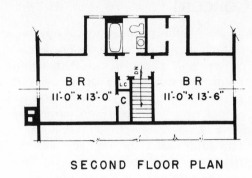

B R
11'-0" x 13'-0"

DN

LC

C

B R
11'-0"x 13'-6"

SECOND FLOOR PLAN

The Fieldstone

Quaint and picturesque in design, this home will be outstanding in any neighborhood. Compact and easy to care for, giving you the most for budget. A vestibule and coat closet offer hospitable welcome as you come in from the covered porch. The living room has two exposures and a door out to the breezeway. The fireplace keynotes a room of distinction.

On the other side, with windows on two sides, too, is the dining room, with an alcove for a built-in china cabinet. The kitchen is sunny, efficient and convenient to the cellar stairs.

At the rear are two bedrooms, featuring cross ventilation and super closets, and the bath between is family-sized. Two more dormered bedrooms can be added in the expansion attic, with the second bathroom utilizing the same plumbing stack of the one below for economy.

AREA: First floor   1,033 sq. ft.
      Second floor   550 sq. ft.

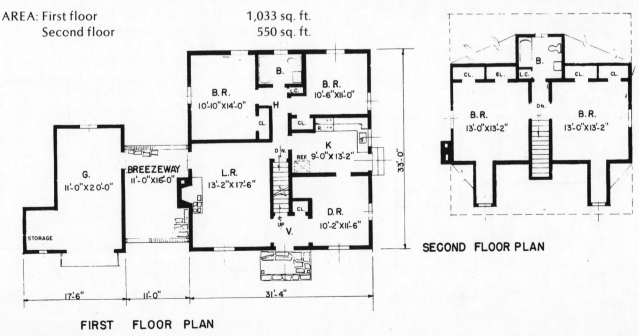

B.

B. R.
10'-10"x14'-0"

H

LC.

B. R.
10'-6"x11'-0"

CL.

CL.

R.

G.
11'-0"x20'-0"

BREEZEWAY
11'-0"x16'-0"

L. R.
13'-2"x17'-6"

DN.

K
REF. 9'-0"x13'-2"

S

33'-0"

STORAGE

UP

CL.

V.

D. R.
10'-2"x11'-6"

17'-6"   11'-0"   31'-4"

FIRST   FLOOR   PLAN

B.

CL.

CL.

LC.

CL.

CL.

B. R.
13'-0"x13'-2"

DN.

B. R.
13'-0"x13'-2"

SECOND FLOOR PLAN

The Concord

This plan is something quite unusual—arranged to give many features not ordinarily obtained in a house with so conventional an appearance. Although we see a typical Cape Cod exterior, the service entrance and kitchen are in the front. In the rear, with large window areas overlooking a very up-to-date terrace, are the living room and dining room which take advantage of a view to the garden area. The two bedrooms and bath on the first floor provide ample space for the small family, and the second floor provides extensive area for expansion into two or more very large bedrooms and another full bath. A breezeway and attached garage complete this plan, which has every convenience for every day living.

AREA: First floor   1,054 sq. ft.
        Second floor   434 sq. ft.

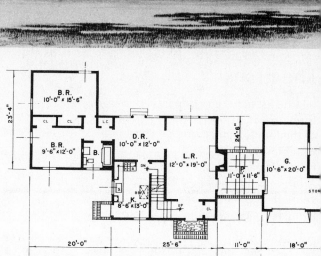

SECOND FLOOR PLAN

FIRST FLOOR PLAN

The Highlander

Designed to meet your every need is a beautiful expansion rambler. It's a perfect house for a 50' lot without the garage, and yet not a thing is missing for lots of living. The overhang at the front door is welcome in wet weather, and inside, the little vestibule with the open stairway up protects the living room as effectively as a full foyer. A convenient guest closet is fitted in very handily, too. The living room is a delight, good-sized and, with a wide picture window at the front and a stunning fireplace, its effect is one of distinction.

The kitchen at the back makes effective use of every inch—a bright work area. The dining alcove enjoys corner window views, and with the side door opening so conveniently to the porch, outdoor dining is effortless fun on pleasant evenings. The two bedrooms can each accommodate twin beds, for a family of four, and when the upstairs is completed, you'll have two or more large chambers, with charming dormer alcoves, a second bath, and a total of eight big closets throughout.

AREA: First floor (excl. garage & porch)   1,077 sq. ft.
        Second floor   735 sq. ft.

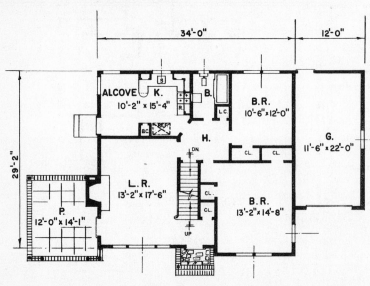

    FIRST FLOOR PLAN

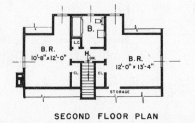

SECOND FLOOR PLAN

SECOND FLOOR PLAN

The Essex

"The Essex" is an outstanding design featuring great flexibility. For older members of the family there is a downstairs bedroom which may also serve as an office for professional people.

A large, fully equipped kitchen, the cross-ventilated dining room and the spacious living room round out the first floor.

Provision has been made upstairs for two large comfortable bedrooms and a complete bath.

AREA: First floor   1,093 sq. ft.
         Second floor   886 sq. ft.

FIRST FLOOR PLAN

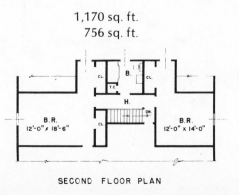

"A cozy little home all our own" is the dream of most families. This is as cozy a home as you'll find, yet it is not condensed down to doll-house size. Waste space has been kept to an absolute minimum without sacrificing convenience and comfort. This plan includes such features as an entrance vestibule, a center-core hall for good circulation, loads of closets, large kitchen with separate dining area and access directly to the garage, basement and porch, not to mention the large living room with an open stair to a future expansion attic. This attic area has space for two more large bedrooms and a full bath.

The Drake

AREA: First floor   1,170 sq. ft.
       Second floor   756 sq. ft.

SECOND FLOOR PLAN

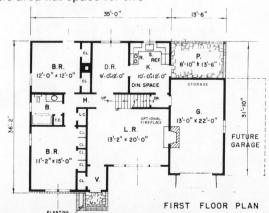

FIRST FLOOR PLAN

The Sherwood

The brick planting bed is shadowed by a projecting gable face which forms wonderful weather protection for the entrance platform. This, combined with high, bright corner windows and a horizontal wood rail fence which separates the breezeway and garage, forms a pleasing exterior. The equally pleasant arrangement of interior plan combines comfort and convenience for perfect living. A full complement of rooms on the first floor, including two bedrooms and bath, are supplemented by an additional two rooms and bath on the second floor. So much living space in so small an appearing exterior is truly amazing. The living room with its specially arranged fireplace location and large corner of window area presents an immediate touch of "something different", which everyone is looking for these days.

AREA: First floor   1104 sq. ft.
       Second floor   570 sq. ft.
       Basement   1110 sq. ft.
       Garage   286 sq. ft.

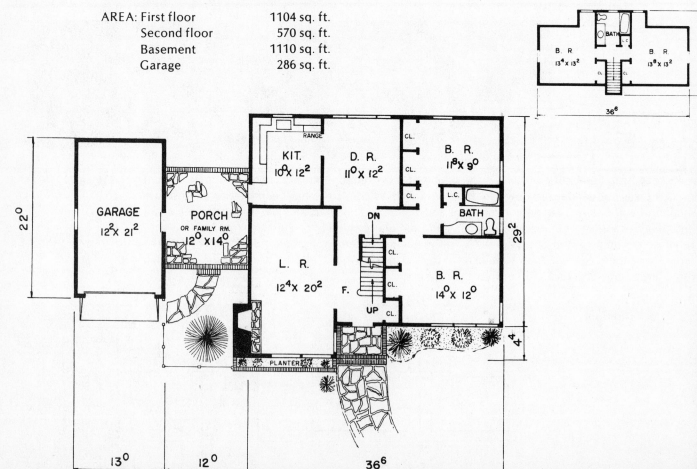

The Cornell

A pleasant combination of the old and the new is expressed in this home's warmth of feeling on the exterior and its convenience of planning on the interior. A separate entrance hall including guest closet and stair to the second floor starts out the plan. As this hall extends back it provides convenient passage to living room, kitchen, basement, bedrooms, and bath respectively. The living room features a full size real fireplace, lots of floor space and entrance to a neat breezeway. The breezeway, in addition to being a wonderful place for relaxing on summer evenings, performs the satisfying function of protecting passage to and from the attached garage in stormy weather.

AREA   First floor   1186 sq. ft.
       Second floor   600 sq. ft.

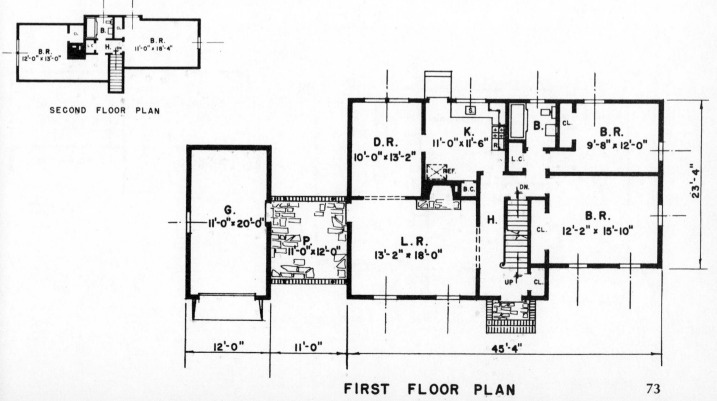

SECOND FLOOR PLAN

FIRST FLOOR PLAN

The Princeton

Here's a simple design that offers complete living on one floor, with an opportunity of utilizing the expansion space upstairs for additional living area, if so desired. The entrance foyer features a dramatic cathedral ceiling. This is a comfortable two bedroom house for a retired couple who no longer require a large home or one of economy and efficiency for a younger couple with one or two children, since it can be expanded from two to four bedrooms as needed.

AREA: First floor   1,300 sq. ft.
        Second floor   600 sq. ft.

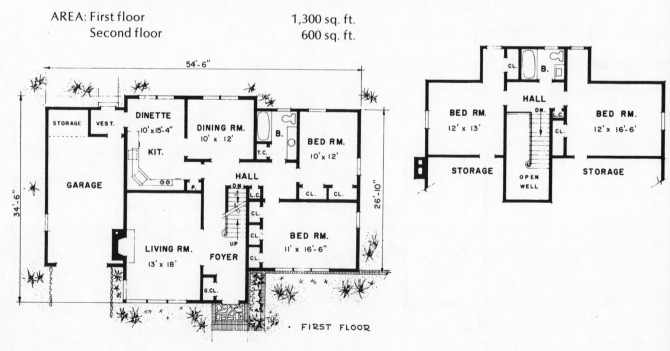

The Allen

All the charm of the traditional Cape Cod house has been recaptured in this story and a half house of fieldstone and siding. Though the house, as it appears in the rendering, is almost 80' long, and perfect for a corner lot, it could be built on a 60' frontage by eliminating the garage and porch, bringing it down to an overall length of 46'. Downstairs, there are two large bedrooms. A bath, well located off the front hall, serves also as a guest lavatory. A spacious, well lighted kitchen makes homecoming a joy. The dining room opens to the porch for cheerful summer suppers. A wide arch joins living and dining areas, and a handsome fireplace suggests unusual and cozy furniture groupings. On the dormered second floor there are two tremendous bedrooms and another full bath, to be finished as you need them, and all bedrooms enjoy two exposures.

AREA: First floor   1,316 sq. ft.
        Second floor   636 sq. ft.

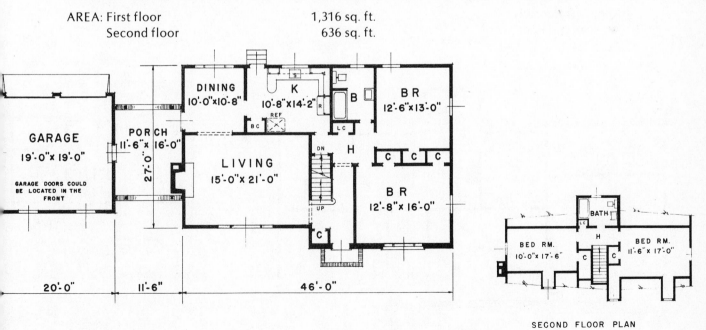

SECOND FLOOR PLAN

The Springbrook

Contemporary planning, combined with time tested advantages, has been included in this won-wonderful design.

Seclusion is to be found in the two large bedrooms, each containing ample closet space. The roomy kitchen takes pride in the sunny dining alcove which will bring added enjoyment to those family meals.

Upstairs two additional bedrooms and a bath are found to serve present or future needs.

Yes, feature upon feature has been included to make this the ideal home for you.

AREA: First floor   1,340 sq. ft.
       Second floor   650 sq. ft.

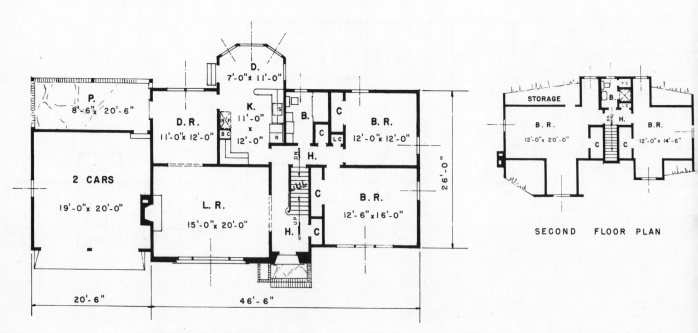

FIRST FLOOR PLAN

SECOND FLOOR PLAN

The Framingham

Quaint and picturesque in design, this home will be outstanding in any neighborhood . . . Compact and easy to care for, giving you the most for your budget . . . A foyer and coat closet offer hospitable welcome as you come in from the covered porch . . . The living room has two exposures and adjoins both the dining room and family room . . . the fireplace keynotes a room of distinction . . . The kitchen-family room is sunny, efficient and convenient to the rear yard . . . Two or three bedrooms, depending on personal requirements, featuring cross ventilation and super closets complete the first floor . . . Two more dormered bedrooms can be added in the expansion attic, with second bathroom utilizing the same plumbing stack of the one below for economy.

AREA: First floor   1,420 sq. ft.
        Second floor   560 sq. ft.

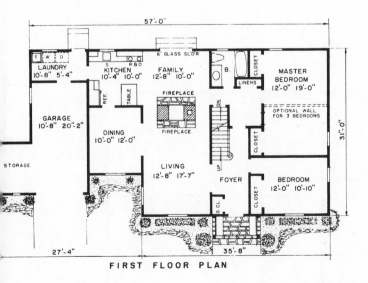

FIRST FLOOR PLAN

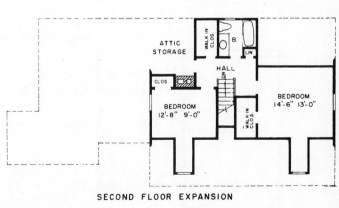

SECOND FLOOR EXPANSION

The Cameron

The luxurious entrance hall featured in this home is the introduction to an over-all plan for wonderful living in modern day style. The recessed entry is flanked by closets on either side, and the open stair lends an immediate impression of beauty and spaciousness. Extending beyond the entry hall is another small hall area which provides direct access to bedrooms, bath and kitchen. Closet space is abundant on both floors. The two rooms on the second floor of this plan are the nicest we have ever seen in a one and one-half story home. There is practically full ceiling height for the entire room area, and there are none of the unsightly breaks which usually occur in these "attic" rooms.

AREA: First floor   1,446 sq. ft.
       Second floor   619 sq. ft.

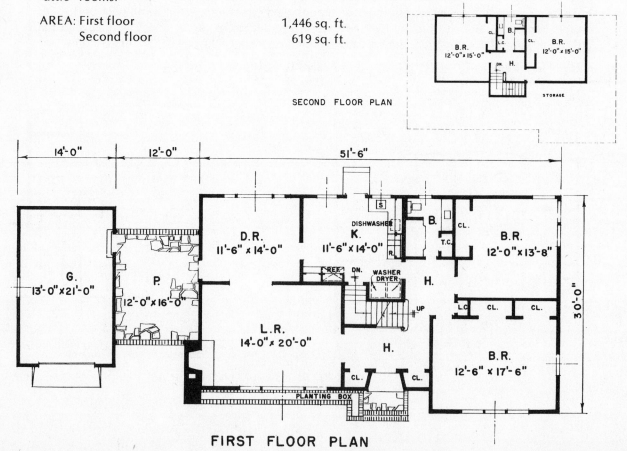

SECOND FLOOR PLAN

FIRST FLOOR PLAN

The Winslow

Here is the ideal home for the growing family.

This home with its overall dimension of 60' can be built on most 75' lots.

Note the desirable location of the family room, directly off the foyer and also adjacent to the dinette. This room, complete with large sliding doors and early American fireplace, will be one of the most used rooms in the house.

Connecting the kitchen and two car garage is a combination laundry-mud room, ideal for mother and children alike.

The living room with its large cottage windows affords maximum wall space for furniture arrangement.

The second floor affords two king size bedrooms, including two oversized closets in each room, and bath. This area can be finished during construction or as the family grows.

AREA: First floor   1,810 sq. ft.
         Second floor   650 sq. ft.

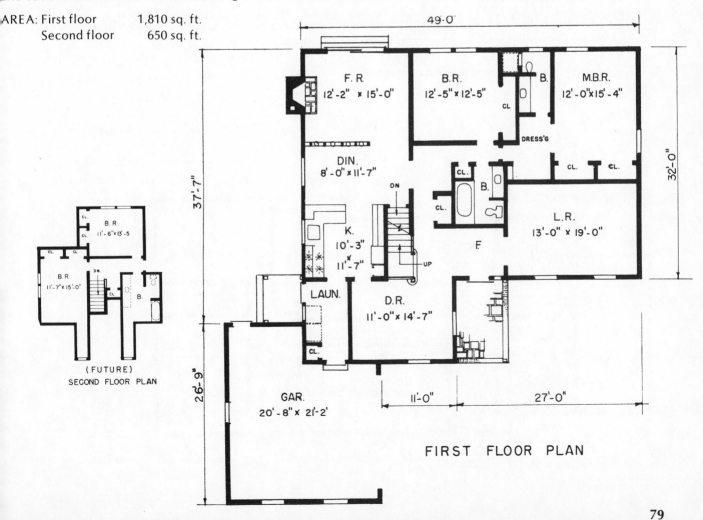

(FUTURE)
SECOND FLOOR PLAN

FIRST FLOOR PLAN

The Taylor

Picturesque colonial living is yours with this three bedroom design. A cozy library off the entrance foyer provides a secluded retreat for the head of the household.

The kitchen dining area is set off with a sunny corner box bay window arrangement.

A two car garage is found in the basement.

TOTAL LIVING AREA: 1,975 sq. ft.

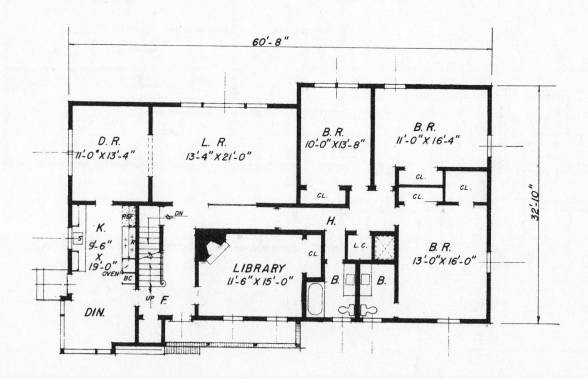

The Wilshire

A flavor of French Provincial architecture is emphasized in this one and one-half story design by the steep roof lines, the massive chimney and the distinctive double entrance doors.

To the right of the spacious foyer is the sunken living room, and to the rear is the wood-paneled family room, which features a corner brick fireplace and a triple sliding glass door that leads to the rear terrace. The U-shaped kitchen and dinette with sliding doors to the rear are a homemaker's delight. A spare room off the laundry and two car garage may be used as a maid's quarters or hobby room.

Two bedrooms, each with its own private bath, complete the first floor.

Upstairs, the two bedrooms and bath which can be finished at a later date, if so desired, are reached by an attractive open-well staircase

AREA: First floor   2,290 sq. ft.
        Second floor   630 sq. ft.
        Garage   420 sq. ft.

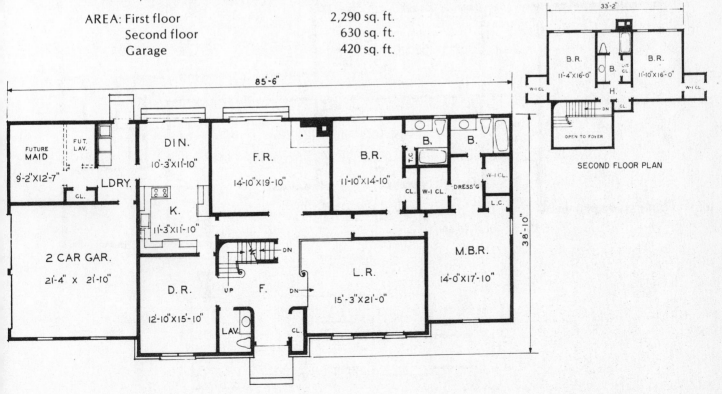

FIRST FLOOR PLAN

SECOND FLOOR PLAN

The Yarborough

The covered front portico exemplifies Colonial grace, together with brick veneer, wood shingles, board and batten vertical siding and shuttered windows to give this one and one-half story house a feeling of timeless beauty. The three car garage features a stair to the basement. Entering the front door is a spacious foyer with twin coat closets. Visible from the foyer is a formal sunken living room with circular bay window and a built-in window seat. The gallery offers a full view of the living room below by means of a knee high wrought iron railing. The traditional fireplace is located in the family room and is "backed-up" by another fireplace that graces the luxury of the master suite.

The attractive open stairway, with wrought iron railing and landing at the half point, leads to the additional three bedrooms on the second floor which may be finished at a later date, if so desired.

AREA: First floor   2,800 sq. ft.
          Second floor   1,130 sq. ft.
          Garage   750 sq. ft.

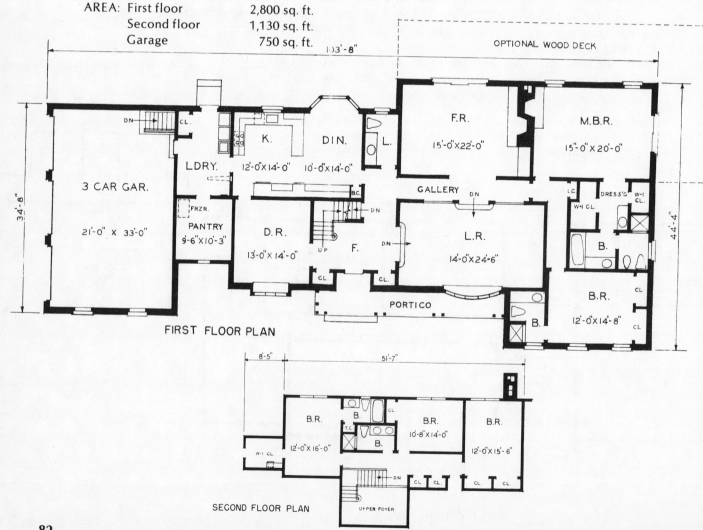

FIRST FLOOR PLAN

SECOND FLOOR PLAN

SPLIT LEVEL

Has three or four levels. Less stair climbing when going from one level to another, but total climbing may be more than in a two-story. Especially suitable for rolling terrain. Lends itself to attractive exterior appearance if well designed. Requires more land than a two-story, but has more liveable space for the money than a ranch.

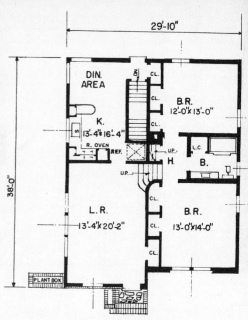

·FIRST & SECOND LEVELS·

The Pitney

Split levels being very popular, almost everyone is considering one type or another for themselves to live in. Perhaps you too would like a split, but you aready have a lot and it is too small for most splits which you have seen. Here is the answer to your problem. This compact yet roomy home is less than 30' in width allowing it to be built on a lot as small as 40' depending on local requirements. There is wonderful closet space in all areas; a large corner window dining space, modern built-in range and oven, snack bar and large utility closet in the kitchen; convenient stair to utility and garage levels and recreation room; an extra hobby room, den or office off the recreation room and a lavatory between for convenience. Two bedrooms and a vanitory bath form the regular bedroom level but there is a stair up to yet another level over the kitchen and living room with space for an extra room with two large closets.

AREA: First & Second levels   1,085 sq. ft.
       Lower level   494 sq. ft.

The Kimberly

Modern planning for convenient and comfortable living is to be found in this contemporary design.

The living and dining areas form a large open area which is highlighted at each end by large bright windows.

Three large bedrooms and a full bath round out the upper levels. A garage, lavatory and a cozy den are to be found on the entrance level.

AREA: First & Second levels   1,110 sq. ft.
       Lower level   300 sq. ft.

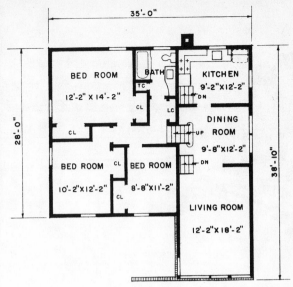

FIRST & SECOND LEVELS

The Hemingway

For a narrow lot this is an ideal split level plan.

Incorporated in only 1116 sq. ft. are six spacious rooms and a bath. In addition, on the lower level, there is a separate entrance foyer leading to a recreation room which opens to the rear patio thru a sliding glass wall. A laundry room and separate lavatory plus an ample sized garage complete this level.

Below in the basement there is a tremendous area which may be put to multi use as an additional play room, workshop, or hobby area.

AREA: First & Second level   1,116 sq. ft.
       Entrance & lower level   305 sq. ft.

BASEMENT & GROUND LEVELS

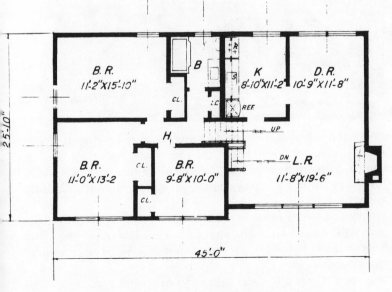

· FLOOR PLAN · FIRST & SECOND LEVELS ·

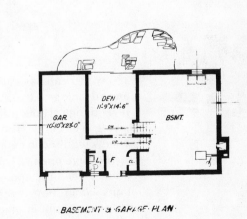

· BASEMENT & GARAGE PLAN ·

85

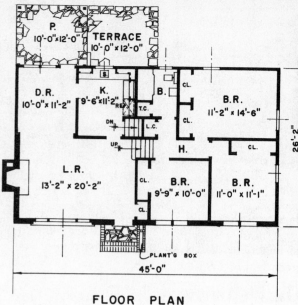

FLOOR PLAN

The Parkway

Split-levels are in the news today. This charming brick and clapboard version is extremely flexible, adaptable to a lot which is flat as well as one with a slope. Here you have a wide open plan for living-dining service, enhanced by a natural brick fireplace wall in the living room and open through the dining room to the bright airy porch with an adjoining terrace at the rear. Six gentle steps lead up to the bedroom area providing all the privacy of a two story home with the convenience of only a half flight of steps. A full size two car garage, an integral part of the house, another half flight down is an important feature for wet weather. There is also a combination laundry-lavatory situated on this garage level, while in the basement proper is ample room for a large recreation area.

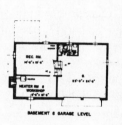

BASEMENT & GARAGE LEVEL

AREA: First & Second levels   1,170 sq. ft.
          Lower level   340 sq. ft.

The Saxony

Now you can have all the advantages of modern planning for luxury and comfort, together with any advantage a two story house might provide, in the economical and increasingly popular split-level design. Three full bedrooms and a bath are located only five steps up from the living area, giving in effect the full privacy afforded by the old two story home, yet saving more than half the energy required in climbing a full flight of stairs.

The large open "L" shape of the living and dining areas in modern open planning at its best and at once gives the appearance of extra size to each area. The kitchen is a wonder of efficiency, all units within short range of each other. Brightly lighted by an extra-sized window, it has plenty of space for a built-in snack bar in its own corner.

As in most split-level designs, the garage is only a few steps down from the kitchen. Located on the same level is an enormous area available for recreation or workshop. This plan also has a full basement area under the living, dining and kitchen areas.

     AREA: First & Second levels   1,190 sq. ft.

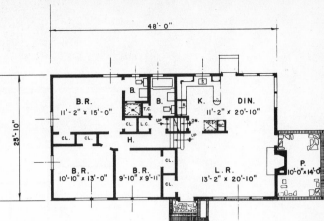

The Normandy

The beauty of a split-level home is threefold. It allows you to fit the house on a sloping lot, saves a great deal of stair climbing by reducing the number of steps by half, and adds to the privacy of living. This split-level home incorporates these advantages with an "extra". The house is so designed that it may also be constructed on level ground. Thus, if you have your eye on — or own — a lot without a slope, you can still build a split-level home.

There are three bedrooms, the master bedroom featuring its own private shower-bath and two separate closets. Speaking of closets, count them: a total of 8 including a broom closet in the kitchen area for holding cleaning equipment that always seems a problem to store.

The kitchen-dinette is a thing of beauty and luxurious convenience with all the counter worktop area anyone could wish for. There is, of course, a service or snack bar. The superlative living room with its real fireplace opens in "L" shape into the dinette, but can be separated by the folding door for table cleaning time. Then, of course, there is the beautiful porch off the living room, a garage under the bedrooms only one-half flight down from the kitchen, and a room for spacious recreation area.

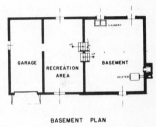

BASEMENT PLAN

AREA: First & Second levels   1,240 sq. ft.
       Lower level   360 sq. ft.

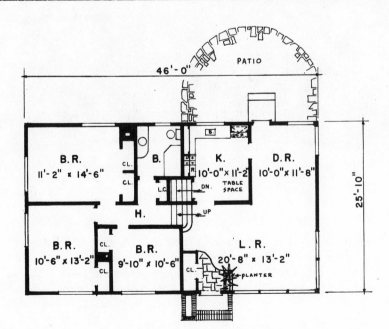

FLOOR PLAN

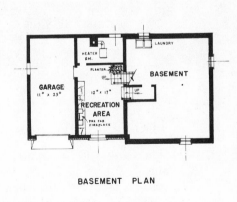

BASEMENT PLAN

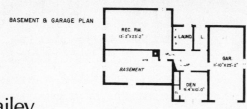

BASEMENT & GARAGE PLAN

REC. RM.
13'-2"X23'-2"

LAUND. L

BASEMENT

GAR.
11'-10"X25'-2"

DEN
9'-4"X10'-0"

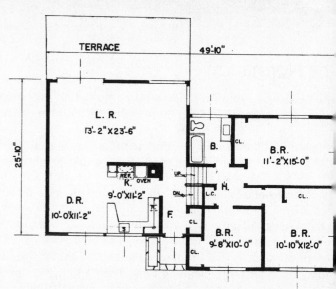

TERRACE   49'-10"

25'-10"

L. R.
13'-2"X23'-6"

B.

CL.

B.R.
11'-2"X15'-0"

K.
9'-0"X11'-2"

OVEN

UP

H.

CL.

D.R.
10'-0"X11'-2"

DN.

F.

L.C.

CL.

B.R.
9'-8"X10'-0"

B.R.
10'-10"X12'-0"

CL.

CL.

The Bailey

A design keeping with the modern trend is yours with this delightful home.

From the extra large windows, and closet space, to the folding partition dividing the kitchen and dining room, this is a home full of those extras that make it outstanding.

At a glance chances are you may find this home to be the one you have always been dreaming about.

AREA: First & Second levels   1,290 sq. ft.
         Lower level   650 sq. ft.

The Bellamy

The foyer, with its connecting garage, offers the ultimate of protection for that bad weather ahead. A sunny den and connecting lavatory bring this level to a wonderful conclusion. An open stairway leads us to the living level with its large living-dining rooms and modern kitchen.

The bedroom levels consist of four, yes four large bedrooms, two baths and all the closets the family can ever use.

AREA: First & Second levels   1,305 sq. ft.
         Lower level   450 sq. ft.

FLOOR PLAN

B.R. 11'-0"x12'-0"

B.R. 11'-2"x15'-2"

B.R. 10'-0"x11'-0"

L.R. 13'-0"x18'-6"

K. 8'-10"x11'-0" OVEN

D.R. 11'-0"x11'-0"

F.

48'-0"

30'-4"

The Oakley

Here is a full size three bedroom, two and one-half bath split level complete with recreation room, garage and full basement—and all this can be built in most areas on a 50' lot.

A most pleasing exterior is obtained in this design in spite of the narrow dimension. A covered entrance platform leads to a true center hall entrance foyer flanked on either side by living room and dining room.

AREA: First & Second levels   1,377 sq. ft.
        Lower level   345 sq. ft.

G. 13'-0"x22'-6"

REC. RM. 14'-11"x21'-4"

B'S'MT.

48'-0"

30'-4"

BASEMENT PLAN

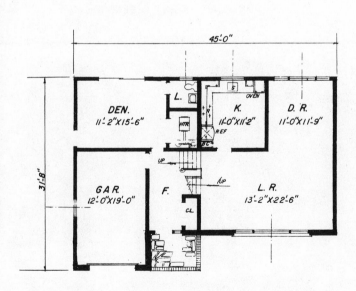

45'-0"

DEN. 11'-2"x15'-6"

K. 11'-0"x11'-2"

D.R. 11'-0"x11'-9"

REF

HTR

GAR. 12'-0"x19'-0"

F.

L.R. 13'-2"x22'-6"

UP

UP

CL

31'-8"

· GROUND & LIVING ROOM LEVELS·

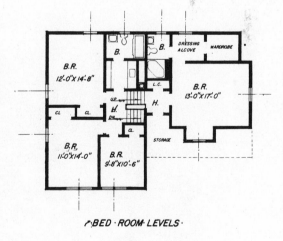

B.R. 12'-0"x14'-8"

DRESSING ALCOVE

WARDROBE

B.R. 13'-0"x17'-0"

CL

H.

STORAGE

B.R. 11'-0"x14'-0"

B.R. 9'-8"x10'-6"

UP

·BED·ROOM·LEVELS·

The Wiley

Three large airy bedrooms line up to overlook the living area from the picturesque balcony above.

In addition to the modern fully equipped kitchen and spacious dining room, this living area boasts a beautiful cathedral ceiling over the living room.

Two full baths, a two car garage and ample closet space round out this smart design.

AREA: First & Second levels   1,390 sq. ft.
         Lower level   365 sq. ft.

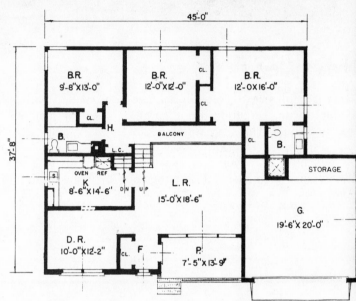

· FLOOR PLAN · · FIRST & SECOND LEVEL ·

BASEMENT PLAN

The Gregory

Created to keep up with the modern trend in house building, this beautiful split-level has been designed to bring you years of happy contented living.

For those relaxing hours the leisurely living room with its cheery hearth and full window is ideal.

Meal preparing and serving will be made easy for Mom in the handy kitchen and sunlite dining room.

Three bedrooms, many closets, and two modern baths are included in the dreamy bedroom level.

AREA: First & Second levels   1,400 sq. ft.
        Third level   300 sq. ft.

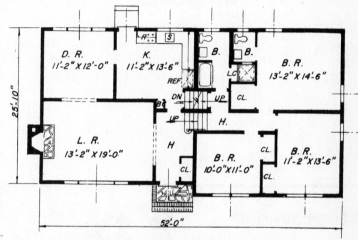

THIRD LEVEL PLAN

The Buckley

Contentment and comfort radiate from this split-level.

Three bedrooms to meet today's needs and a fourth on the third level for future use all are to be found carefully included in this design.

The lower level contains a garage for the family car and a relaxing recreation room.

AREA: First & Second level   1,400 sq. ft.
        Third level   275 sq. ft.
        Lower level   400 sq. ft.

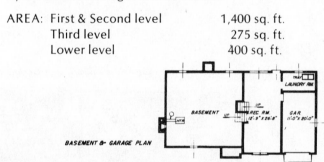

BASEMENT & GARAGE PLAN

FLOOR PLAN FIRST & SECOND LEVELS

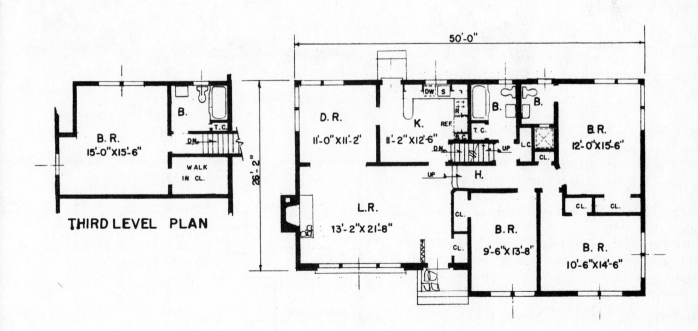

THIRD LEVEL PLAN

The Lowery

Every inch of space has been put in use to bring you a beautiful roomy split-level home for those prospective home builders with limited lots.

The sleeping level includes three spacious bedrooms and two full baths. For just plain homey living the lower level is highlighted by the large living room, dinette, and fully equipped kitchen.

Below, the family recreation room and two car garage round out a home that has soared to popularity.

AREA: First & Second levels   1,406 sq. ft.
         Third level   215 sq. ft.
         Lower level   220 sq. ft.

The Falmouth

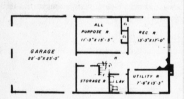

BASEMENT & GARAGE PLAN

The living area of this "Ranch-Split" design is all on one floor. The unusual feature of the "split" entrance vestibule creates an exterior appearance of the formal "Two story colonial." This entrance feature also makes available the entire basement area for finished liveable rooms. The basement being only partially below grade in the front and on grade in the rear.

This ingenious arrangement of making available the finished living areas in the basement tends to put this home nearer the category of the 2 story home for economy of construction based on the total square feet of living area.

First Floor Living Area: 1430 sq. ft.
Lower level   832 sq. ft.

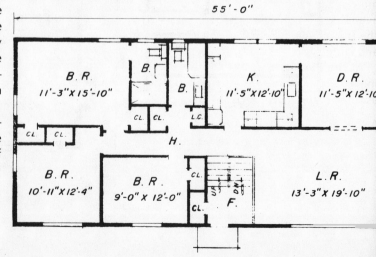

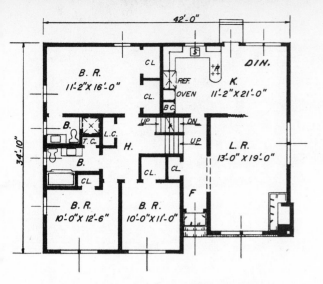

FLOOR PLAN FIRST & SECOND LEVELS

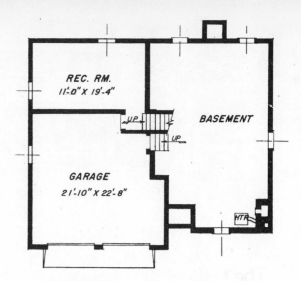

BASEMENT & GARAGE PLAN

The Copley

The typical New England colonial farm house styling of this home contains behind its homey exterior a surprising arrangement of rooms evolved by present-day architecture. It will give you a most convenient and comfortable living experience at the most economical cost.

Basically a three-bedroom, two-bath plan; because of the multi-level design there is a wonderful recreation room available with full-size windows just a few steps down from the living level. Convenient at this level also is a lavatory and laundry area and a full two-car garage.

At the next level above the bedrooms, and over the living area is an expanded attic area large enough for as many as three additional rooms and a bath.

AREA: First & Second levels   1,487 sq. ft.
       Lower level   420 sq. ft.

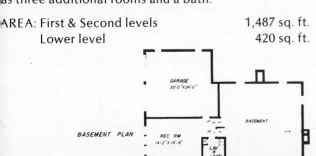

BASEMENT PLAN

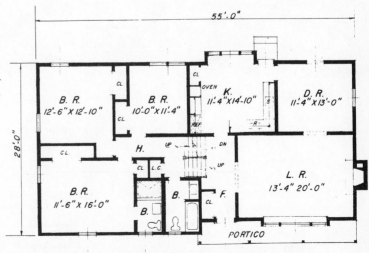

·FLOOR PLAN · FIRST & SECOND LEVELS·

The Emery

"Something different" keynotes the planning that is to be found in this split design.

The lower level is formed by a two car garage, Mom's laundry, and the family room. A few steps up and we find the kitchen and sunny living room which run across the rear of the house.

Going up a few more risers we have three well lighted bedrooms and two baths.

AREA: First & Second levels   1,493 sq. ft.
           Lower level   705 sq. ft.

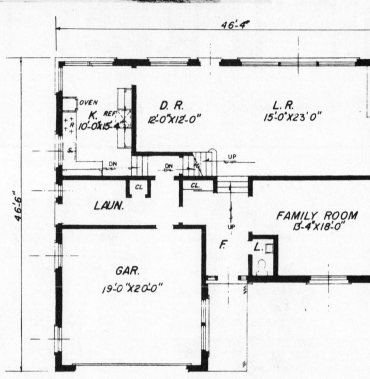

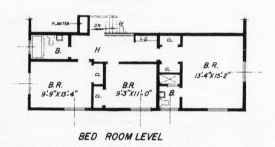

BED ROOM LEVEL

The Dudley

Found to be one of the most desirable and economical split-level desings, this home has the added grace of taking up a relatively small area to build.

In addition to the lower levels which are designed to utilize every inch of space, another bedroom has been planned into the top level to be finished at your convenience.

AREA: First & Second levels   1,512 sq. ft.
           Third level   370 sq. ft.
           Lower level   290 sq. ft.

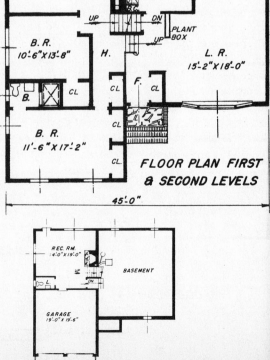

FLOOR PLAN FIRST & SECOND LEVELS

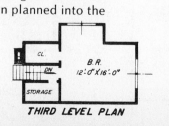

THIRD LEVEL PLAN

The Harmony

The contemporary styling of this multi-level home portrays the up-to-date living comfort which the planning provides.

The on-grade entrance gives a ground hugging character not usually found in split-level homes. Notice how the main roof eases down over the entrance and across the panoramic window of the recreation room to cover the garage entrance, thereby providing weather protection across the entire projecting front of the house.

A full complement of three bedrooms and two baths make up the sleeping level. The living level contains an extremely spacious kitchen with a projecting bay for dining.

The elaborate recreation room has front and rear exposure with sliding doors at the rear to a patio area. Nestled behind the garage and directly adjoining the rear yard is a spacious laundry and mud room.

AREA: First & Second levels   1,498 sq. ft.
         Recreation & Laundry level   510 sq. ft.

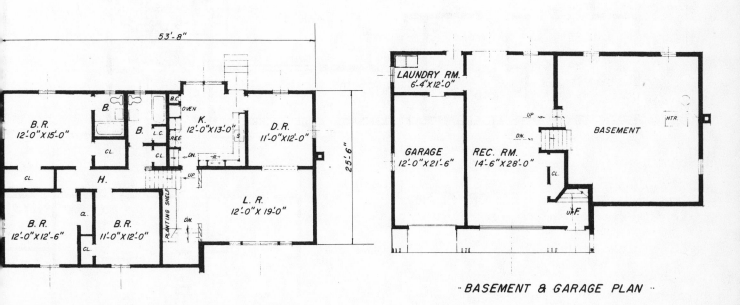

· FLOOR PLAN · FIRST & SECOND LEVELS ·

· BASEMENT & GARAGE PLAN ·

The Pomeroy

Typical of many split-level homes, this delightful house has the convenience of an all-on-one-floor design combined with the privacy that only a two-story can provide.

The bedroom level is made up of three spacious bedrooms, with large closets and two baths, one of which includes a dressing area.

Everyone will love the living area, which includes a carefully planned kitchen, dining room and a friendly living room.

Highlighting this design for modern living, there are two roomy porches and an adjoining garage.

AREA: First & Second levels   1,520 sq. ft.
        Lower level   832 sq. ft.

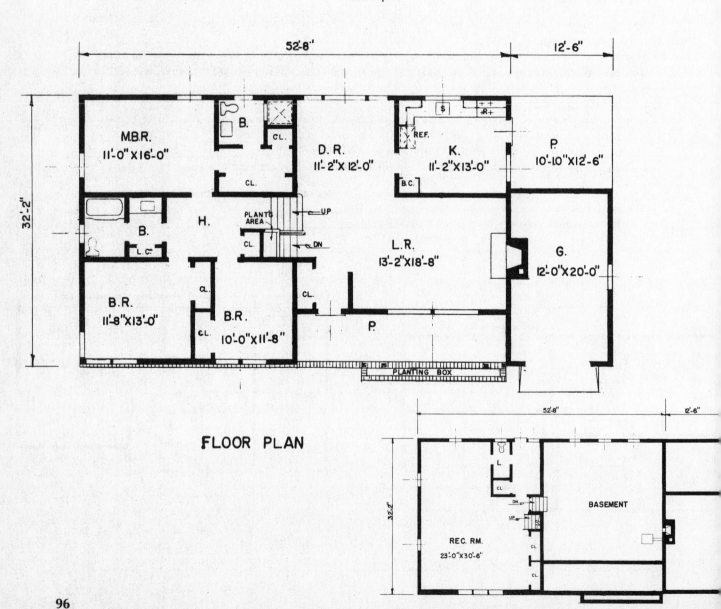

FLOOR PLAN

BASEMENT PLAN

The Conway

The compact arrangement of this split-level plan provides a center hall plus a full sized living room and a attached porch in an overall size no larger than normal. Although this plan does not show a garage—that level being devoted to an enormous recreation area and separate laundry—you may choose to add one to the porch if your lot is large enough. Or you can build a detached garage away from the dwelling. You can even provide for garage doors under the bedroom and still reserve sufficient area for a small play room.

AREA: First & Second levels   1,566 sq. ft.
          Lower level   800 sq. ft.

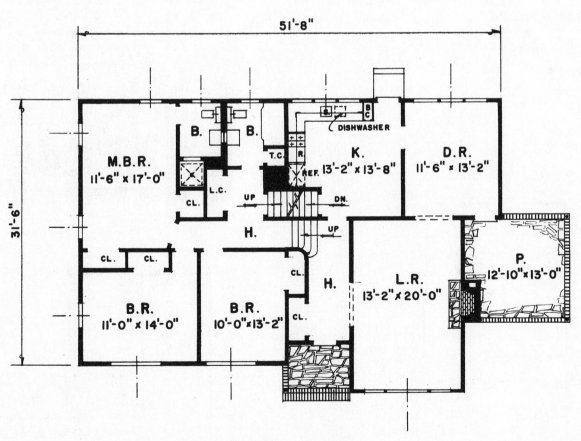

FLOOR PLAN

The Sterling

There is no doubt that a split-level house offers considerable extra space at a minimum of building cost, and although the living, bedroom and recreation rooms are on different levels, the number of steps between levels does away with a lot of stair climbing at a single time.

The first thing that one notices in this three bedroom split-level is the interesting effect created by the sweeping rooflines and the transomed window treatment in a pleasant vertical style, intermixed with brick veneer and vertical siding.

For those who can orient themselves to "split-level" living this carefully planned design, with a long list of features to recommend it, offers visual satisfaction and comfortable living for everyone.

AREA: Upper levels   1278 sq. ft.
        Lower level   288 sq. ft.
        Basement   628 sq. ft.
        Garage   264 sq. ft.
        Deck   252 sq. ft.

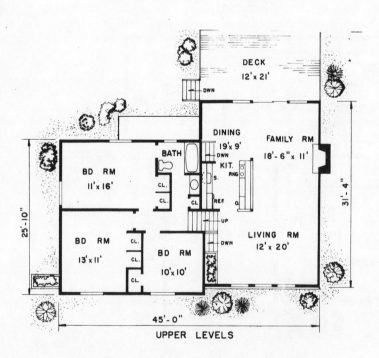

UPPER LEVELS

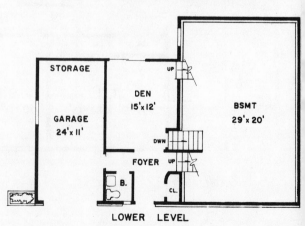

LOWER LEVEL

The Ripley

An elegant exterior complements the outstanding internal planning of this regal home.

As we pass through the foyer, which is flanked by the recreation room, we climb a few risers to the living and dining rooms with an adjoining kitchen.

The upper levels are made up of four bedrooms, two full baths plus spacious closet and storage room.

AREA: First & Second levels   1,260 sq. ft.
        Third level   310 sq. ft.

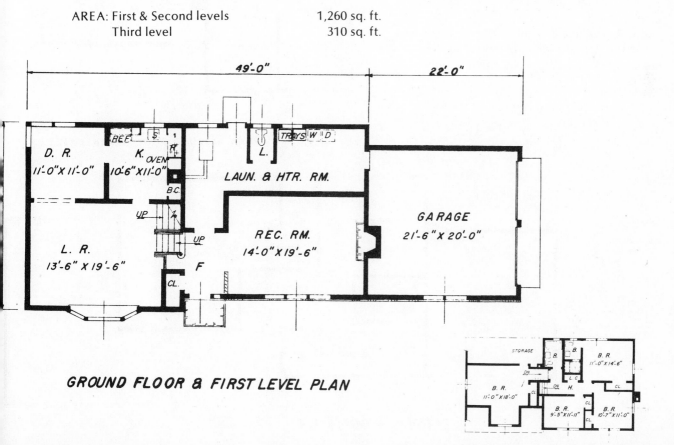

GROUND FLOOR & FIRST LEVEL PLAN

SECOND & THIRD LEVEL

The Academy

Having the outward long look of a ranch, we are immediately aware as we enter of the fact that a modern split-level is to be found within.

Five welcoming steps take us from the foyer to the large living room and from there only eight more take us up to the family slumber area. Located below are a maid's room, recreation area and a two car garage.

In a design like this only a careful analysis of the actual plans will reveal all that has been included to make this the home for you.

AREA: First & Second levels   1585 sq. ft.
         Lower level   825 sq. ft.

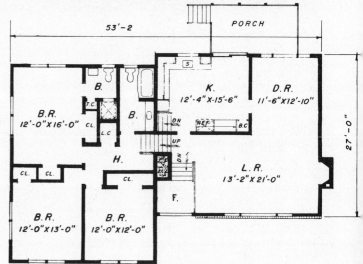

FLOOR PLAN — LIVING & SLEEPING LEVELS

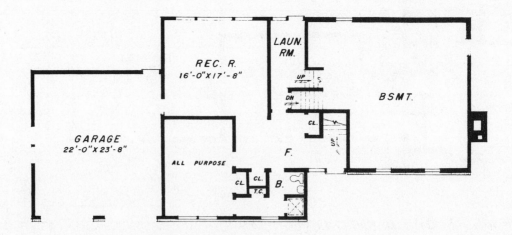

BASEMENT & GARAGE PLAN

The Woodley

Five levels, and all fully usable, make this moderate sized multi-level home chock full of living. A fully excavated basement provides heater, laundry and workshop areas. The second level is occupied by an extra deep 2 car garage, a recreation room and lavatory. Next comes the living level with a thru-hall, extended kitchen breakfast area, a full dining room, living room and porch.

A few steps up from here we then have three full bedrooms with plenty of closets and 2 baths. Then, to top everything off, there is room on the top level for a future room and bath by the addition of a dormer front and rear.

AREA: First & Second levels   1,638 sq. ft.
          Lower level   300 sq. ft.

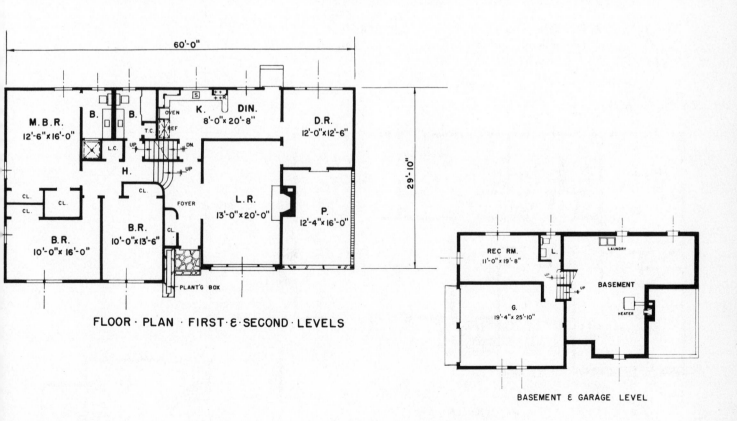

FLOOR · PLAN · FIRST · & · SECOND · LEVELS

BASEMENT & GARAGE LEVEL

The Kelsey

A popular model with a slightly different twist, this split-level offers comfortable living with many luxury features in an economical plan.

Only 2 steps from living to bedroom areas creates the separated effect and practically eliminates stair climbing as such. The full size 2 car garage still leaves room on this intermediate level for a good sized recreation room and a lavatory. The working drawings show an alternate for locating garage doors in the side or front. There is a stair up to the attic which offers convenient access for storage in this large area.

AREA: First & Second levels   1,645 sq. ft.
        Lower level   425 sq. ft.

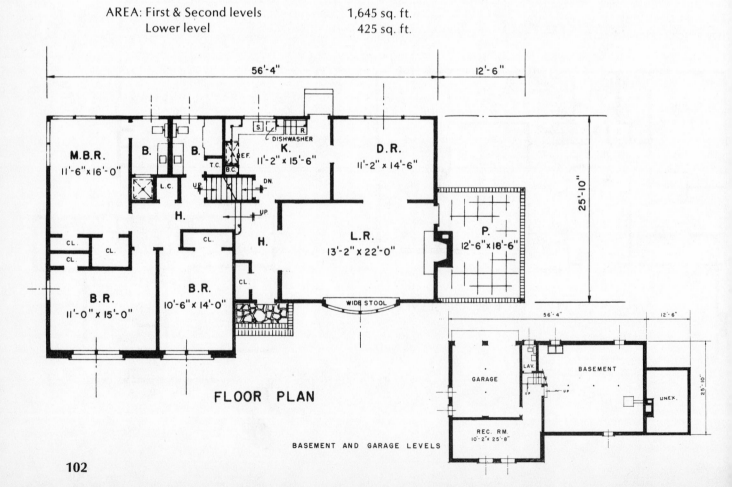

The Rawley

This split-level beauty has been designed for those who want the finest in house planning.

The slumber level comprises three spacious bedrooms and two handy baths.

A welcoming atmosphere prevails in the living level with its extra large living room, dining room, and roomy kitchen.

The lower level contains a two car garage and the family recreation room.

AREA: First & Second levels   1,650 sq. ft.
       Lower level   351 sq. ft.

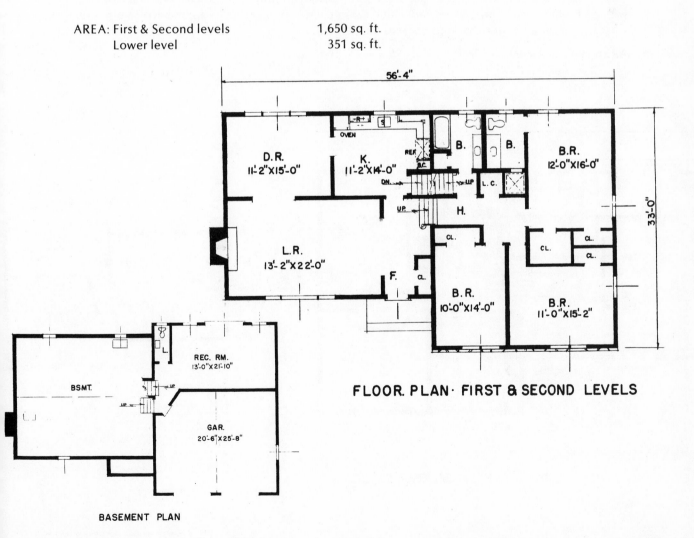

FLOOR PLAN · FIRST & SECOND LEVELS

BASEMENT PLAN

The Bentley

Graceful lines and proportions highlight this exciting split-level. Inside we find bright, cheery rooms, each well designed for the maximum of utility and comfort. The modern kitchen is typical with its built-in oven, counter top range and those useful cabinets Mom appreciates. No matter what room is your favorite habitat, you will find it ideal.

AREA: First & Second levels   1,650 sq. ft.
       Lower level   338 sq. ft.

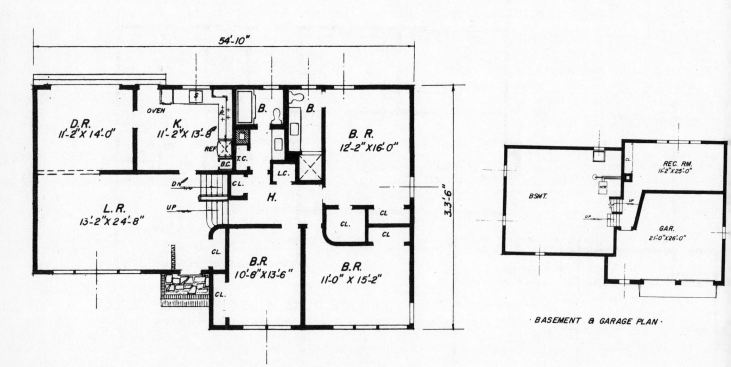

· FLOOR PLAN · FIRST & SECOND LEVELS ·

· BASEMENT & GARAGE PLAN ·

The Argosy

Here is the "different" split level house that you've been looking for. Clean crisp design and contrasting exterior accented by deep shadows and planting beds mark its individual character.

Entering the smart recessed split vestibule, an open living room with fireplace invites you to the first level complete with dining room and large kitchen. Up four stairs are four large bedrooms separated by banks of closets, and two baths assure quiet and privacy.

Downstairs, the recreation room provides many hours of relaxed indoor and outdoor living thru the glass doors. Note the convenience of the kitchen stairs and the large basement for storage.

AREA: First & Second levels   1,665 sq. ft.
        Lower level   335 sq. ft.

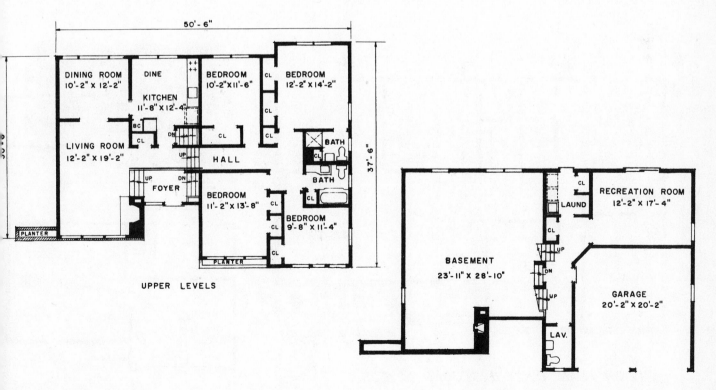

UPPER LEVELS

BASEMENT & LOWER LEVEL

The Roxy

As modern as tomorrow is this breathtaking contemporary design. The exterior is a combination of those extra touches which make a home outstanding in any neighborhood.

Inside we find a wealth of features which add beauty and pleasure to our hours at home. The living room is two short steps down from the partially flagstoned foyer. Tapered slats, a cheery raised hearth, and the breathtaking corner window bring a contented friendly atmosphere to the living room. The carefully designed dining room is two risers above the living room and is on the same level as the roomy, fully equipped kitchen. The bedroom area, a few steps above and beyond the living room, consists of three bedrooms, many closets and two full baths.

AREA: First & Second levels   1,670 sq. ft.
         Lower level   350 sq. ft.

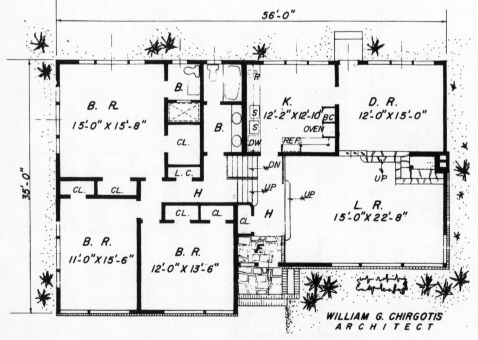

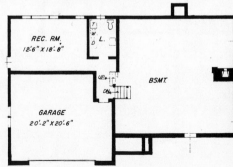

WILLIAM G. CHIRGOTIS
ARCHITECT

·FLOOR PLAN· FIRST & SECOND LEVELS·

·BASEMENT & GARAGE PLAN·

The Finley

Three large bedrooms and two baths overlook the living area, which is just a few short steps below.

Here we behold modern comfort skillfully planned into the friendly living room, sunny dining room, and convenient kitchen.

Ample provision has been made in the basement level for work and recreation rooms.

AREA: First & Second levels   1,670 sq. ft.
       Future Third level   237 sq. ft.
       Lower level   660 sq. ft.

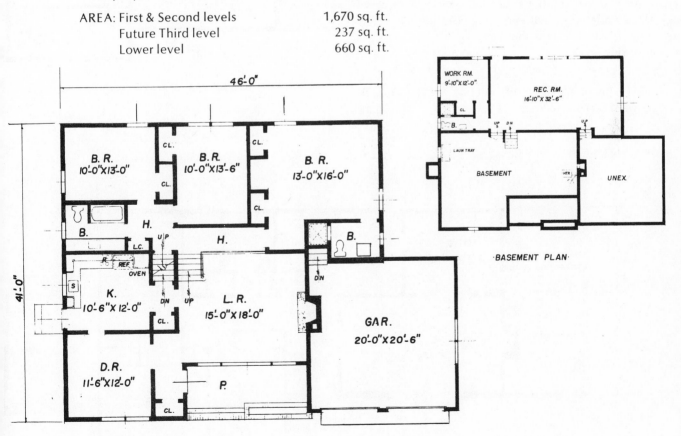

· BASEMENT PLAN ·

· FLOOR PLAN · FIRST & SECOND LEVELS ·

The Westley

A new idea with an old look. If you dream of the charm of the Cape Cod style, yet hesitate because you like the convenience of the split-level plan, here is the home for you. All the features of split-level living are enclosed in this charming Cape Cod exterior. Tremendous kitchen, thru-hall entrance, large living room and dining room, three full bedrooms and two baths plus an extra bedroom and large storage room on the third level over the living area.

There are a large recreation room and lavatory plus a two care garage under the bedroom wing. The garage enters from the rear, but the plans show an alternate for doors in the front or side.

AREA: First & Second levels (excl. porch)   1,700 sq. ft.
           Lower level   400 sq. ft.

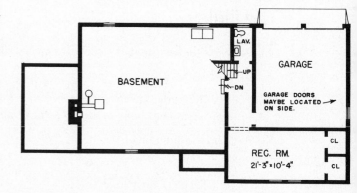

BASEMENT & GARAGE LEVELS

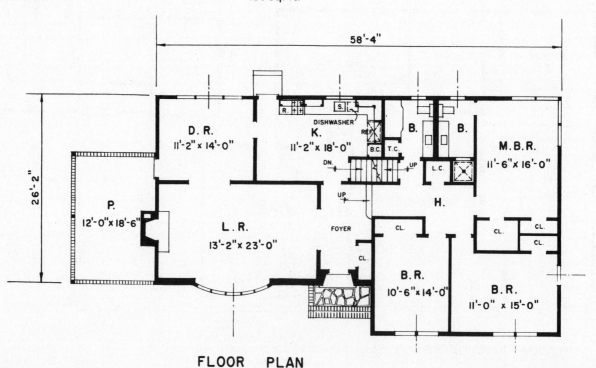

FLOOR PLAN

The Brookway

A split-level with an attached garage? Of course! If your family likes to spend most of its time in the recreation room, here is the plan for you. This many purpose room has been brought up out of the basement, enlarged and opened up to the entrance hall. Planned for the large family, this home has an extra room on the recreation area level with a semi-private shower bath for guests or maid. There is also an open stair to the 3rd level which can be finished off into another large bedroom, bringing the total up to 5 bedrooms. This home for you and your family has "everything" for modern day living convenience.

AREA: First & Second levels   1,715 sq. ft.
(excl. porch & garage)
      Lower level   300 sq. ft.
      Future third level   455 sq. ft.

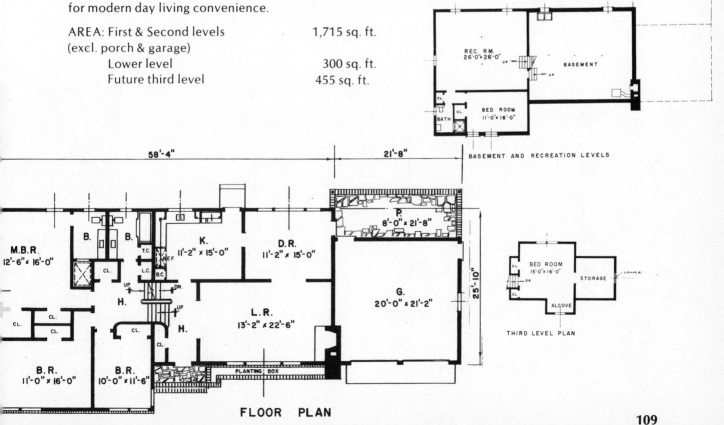

FLOOR PLAN

The Bradley

The most unique feature of this luxurious new split-level home is the adaptation of seven major gas appliances for economical, carefree convenience.

The basement level laundry room features a gas-operated clothes dryer, while a gas furnace, gas hot water heater and gas air-conditioning unit in the utility section maintain year 'round comfort throughout the house. A completely finished recreation room, lavatory, and two-car garage complete the basement-garage level.

The efficiently designed kitchen contains a gas range and gas refrigerator, spacious work counter and abundant cabinet space. The kitchen commands easy access to the dining room, which overlooks grounds and gardens from two directions. A graceful arch leads to the living room, where a 10 foot picture window dominates the front wall. A cozy fireplace is centered in the side wall of this well-proportioned room.

The sleeping quarters are separated from the living room and service areas by a short flight of steps. Sliding doors on the roomy closets add easily utilized space in each of the three large bedrooms. Two full bathrooms complete the bedroom wing.

AREA: First & Second levels   1,715 sq. ft.
        Lower level   450 sq. ft.

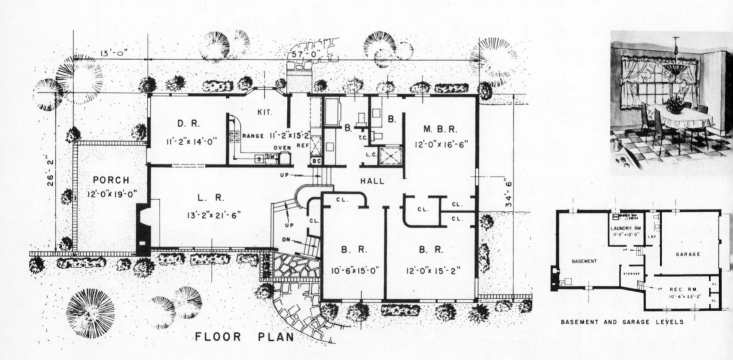

FLOOR PLAN

BASEMENT AND GARAGE LEVELS

The Gray

This exceedingly conventional colonial exterior clothes a very up-to-date and dramatic interior. Split level in arrangement, yet with an unusual flair. An entrance on the ground level opens into a spacious foyer and hall leading on one side to a recreation room and lavatory, on the other to the garage and at the end on a few steps down to the basement and up to the living area. The almost 38' spread of the living-dining room opens out to the rear through two large bay windows, and the high ceiling provides a balcony effect from the bedroom hall. There is a separate service stair leading directly from the kitchen to the bedrooms, and all the bedrooms have good privacy for access to the bath, the master bedroom having its own private full bath.

AREA: First & Second levels   1,725 sq. ft.
         Lower level   500 sq. ft.

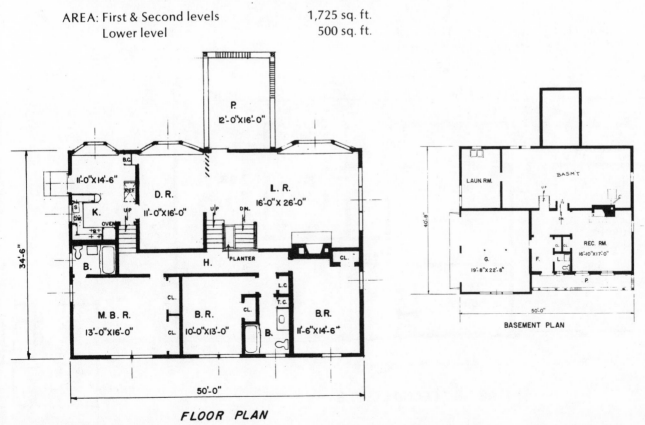

FLOOR PLAN

BASEMENT PLAN

The Rahway

Entrance portico, narrow clapboards, brick, and diamond paned windows are carefully blended to bring you this attractive split level design.

A spacious foyer forms a most impressive reception area which leads to a balconied living room with large picture window and colonial fireplace.

The kitchen with its picturesque curved bay window overlooking the rear garden is only one blessing found in this design.

The lavatory adjacent to the kitchen and recreation room is located for family, as well as guest use. The glass sliding doors of the recreation room take full advantage of view, light and access to the rear yard, patio and garden areas.

Also, this plan provides a den with its own closet directly off the foyer, which can also be used as an overnight guest room.

There are ample sleeping accommodations with three spacious bedrooms, each having sufficient closet space. Also on this same level you will find two lovely baths including a stall shower for the master bedroom bath.

The oversized garage with incorporated storage area completes this split level design.

AREA: Living level   1745 sq. ft.
          Bedroom level   672 sq. ft.

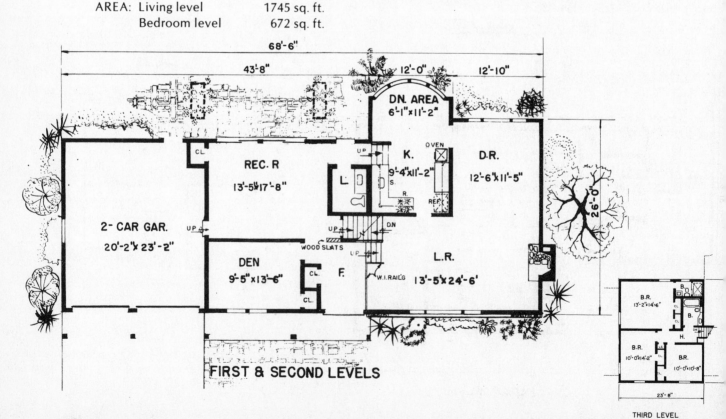

FIRST & SECOND LEVELS

THIRD LEVEL

The Hillery

This is really Colonial split-level living at its finest. The best features of the traditional and the contemporary have been blended to bring you an outstanding home.

For those slumber hours you will appreciate the bedrooms being five short steps up and away from the living room. Everywhere we look we find those wonderful extras like big window areas, closets, etc., which we all desire for that dream house we hope to build.

AREA: First & Second levels   1,770 sq. ft.
        Lower levels   750 sq. ft.

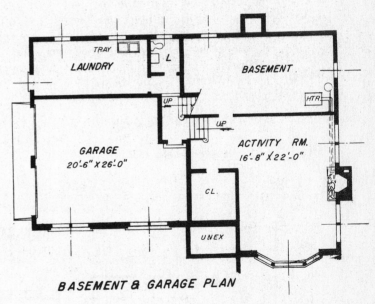

BASEMENT & GARAGE PLAN

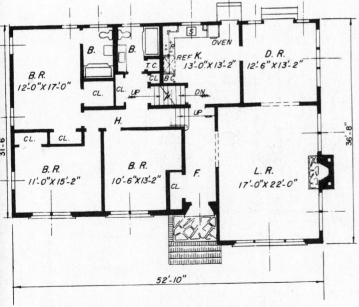

FLOOR PLAN FIRST & SECOND LEVELS

The Surrey

A new look and some improvements on an old favorite go to make up this intriguing split-level home. The charming lines of brickwork extending across and beyond the front to form a lamp-post wall give this dwelling the down to earth feeling of growing out of the ground and blending with the slope of land rather than that of a house set on a lot. The long bow-bay window enhances the front as well as broadening the effective size of the living room itself. The kitchen is a dream in efficiency, lots of work counter and a separate eating area at a large window. Two steps form the split, which practically brings this home down to the ranch level, yet still provides that slight difference between sleeping and living areas. The garage as shown here enters from the rear of the house, but, depending on your requirements, it could be on the side, and the plans show an alternate for this arrangement.

AREA: First & Second levels   1,800 sq. ft.
(excluding porch)
Lower level   255 sq. ft.

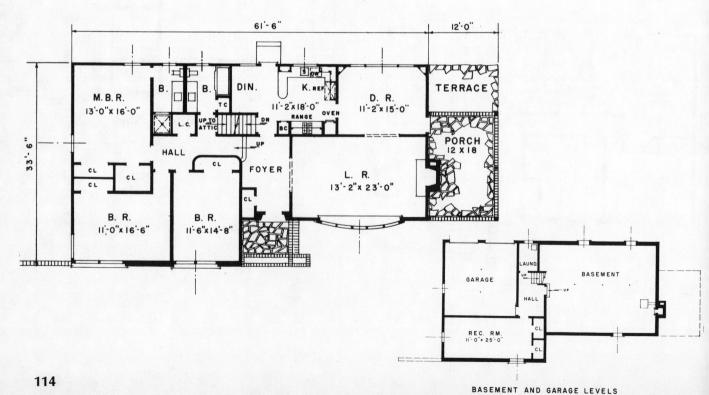

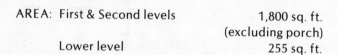

114

BASEMENT AND GARAGE LEVELS

The Berkley

The dramatic exterior appearance of this home carries throughout the interior. Immediately upon entering there is a large roomy foyer with open steps up to the bedroom wing and another short flight to the extra 3rd level bedroom.

Ahead through the arch is the living room, which is separated unobtrusively by planters from the dining room on the right giving an unbroken expanse of 35 feet of space.

The kitchen across the front, almost 24 feet long, includes all of the latest features plus a brightly lighted corner-window breakfast alcove.

The bedroom level features many large closets (walk-ins for the master bedroom) and two full baths, one master — private and the other convenient to the living level to serve as a powder room for guests.

Down ½ flight from this same foyer is a large 2 car garage plus a recreation room, lavatory and laundry. The garage doors here enter from the rear, but with slight adjustments by your builder they could be located on the side or front.

AREA: First & Second levels   1,800 sq. ft.
         Lower level   310 sq. ft.

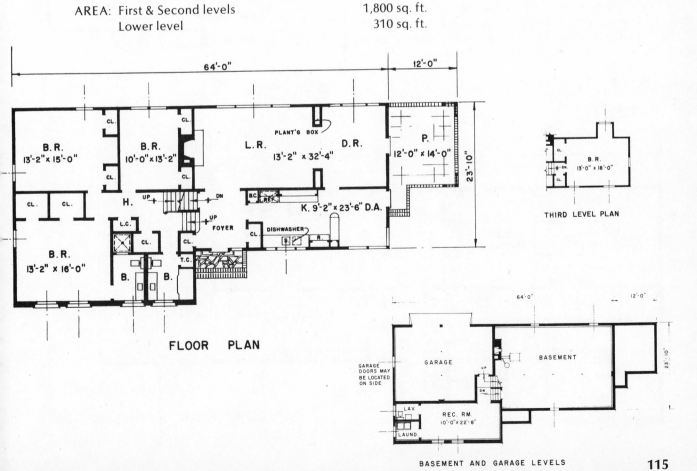

FLOOR PLAN

THIRD LEVEL PLAN

BASEMENT AND GARAGE LEVELS

The Westbury

Did you say closets? Well, here they are! Two closets in every room and all full size—those in the front bedroom are extra large walk-in type. There's a linen closet in the hall, an extra storage closet, a towel closet in the bath, a guest closet in the entry and a broom closet in the kitchen. Eleven closets, and that's not all. The lower level contains an enormous recreation room, a lavatory, laundry and a large storage closet. Everything is large in this home—lots of kitchen with a separate breakfast alcove—a tremendous expanse of living-dining area—a beautiful porch off the side. It's topped off by three of the most spacious bedrooms anyone could want and two full bathrooms.

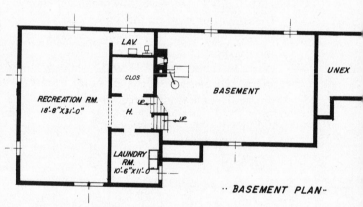

·· BASEMENT PLAN ··

AREA: First & Second levels   1,811 sq. ft.
        Lower level   500 sq. ft.

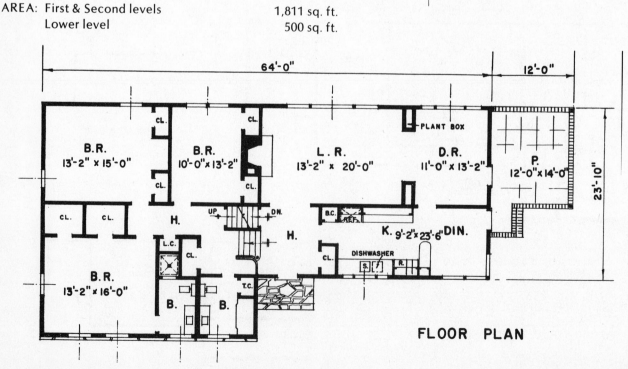

FLOOR PLAN

The Montgomery

Wrought iron posts and railings, brick, and shutters are carefully blended to bring you this attractive split-level design. There are ample sleeping accommodations with three spacious bedrooms, each having sufficient closet space. Provision has also been made for a fourth bedroom if needed. The entrance level provides us with a two car garage, laundry and powder rooms, and a multi-purpose room.

AREA: First & Second levels   1,820 sq. ft.
        Third level   225 sq. ft.
        Entrance level   490 sq. ft.

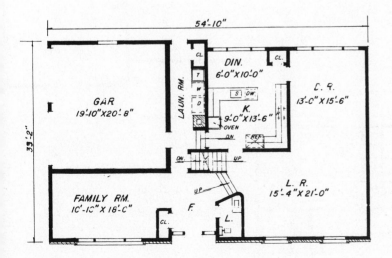

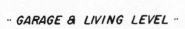

·· GARAGE & LIVING LEVEL ··

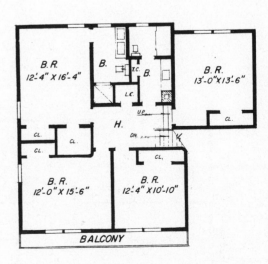

··BED ROOM LEVEL··

117

The Morley

Elegant living for you and your family is yours with this modern split-level home. The living level has a cozy sunken living room. Convenience has been planned into the kitchen's design which also includes a dining area. The dining room with its spacious picture window and adjoining porch is ideal for entertaining those special guests. The secluded bedroom level consists of three charming bedrooms, two roomy bathrooms and all the closet space you'll ever need. A family recreation room with connecting lavatory is found in the lower level next to a full two car garage.

AREA: First & Second levels   1,830 sq. ft.
        Lower level   350 sq. ft.

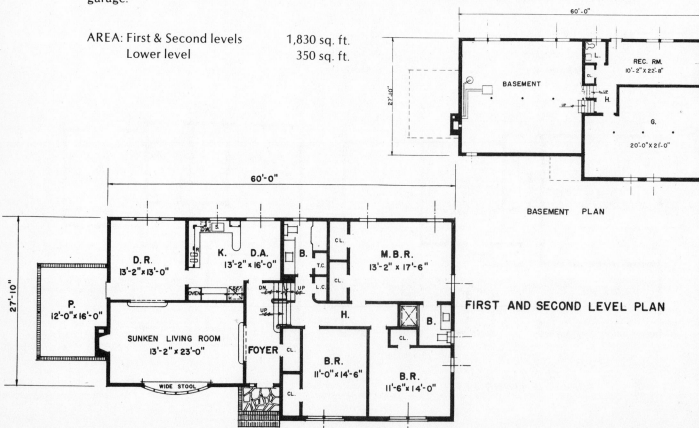

BASEMENT PLAN

FIRST AND SECOND LEVEL PLAN

The Crosby

Having the outward long look of a ranch, we are immediately aware as we enter of the fact that a modern split-level is to be found within.

Two welcoming steps take us from the foyer to the large living room and from there only four more take us up to the family slumber area.

Located below are a maid's room, recreation area and a two car garage.

In a design like this only a careful analysis of the actual plans will reveal all that has been included to make this home for you.

AREA: First & Second levels   1,845 sq. ft.
          Lower level   440 sq. ft.

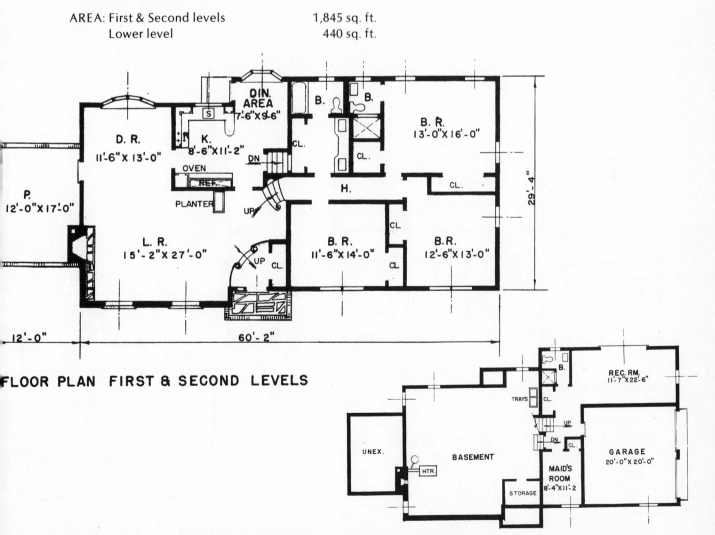

FLOOR PLAN FIRST & SECOND LEVELS

BASEMENT & GARAGE PLAN

The Shawney

Custom designed to fit the requirements of a family living in the metropolitan suburbs, this plan retains the comfort of colonial tradition in its exterior appearance yet provides all the modern facilities for better living.

Large spacious rooms, ideally located for minimum walking distances and well screened and separated for privacy and quiet, combine to give you an end result of architectural perfection for the individual family.

AREA: First & Second levels   1,895 sq. ft.
          Lower level   482 sq. ft.

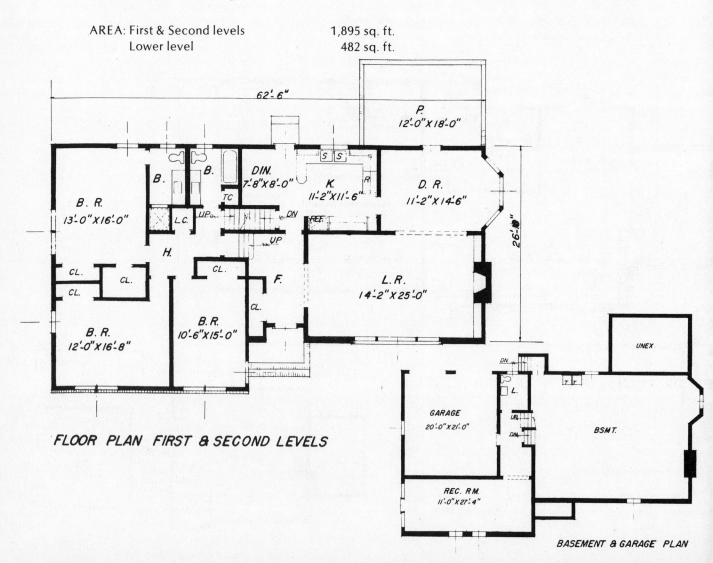

FLOOR PLAN FIRST & SECOND LEVELS

BASEMENT & GARAGE PLAN

The Ainsley

Traditional exterior design gives a warm appearance to this three bedroom split-level design, with the right side of the house featuring wood shingles, diamond paned windows and scalloped gable. Brick veneer is used on the front with softness supplied by covered portico and wrought iron arches. The excellent first impression created by the long attractive porch is carried past the entrance door into the spacious interior.

AREA: First & Second levels   1,900 sq. ft.
        Lower level   308 sq. ft.

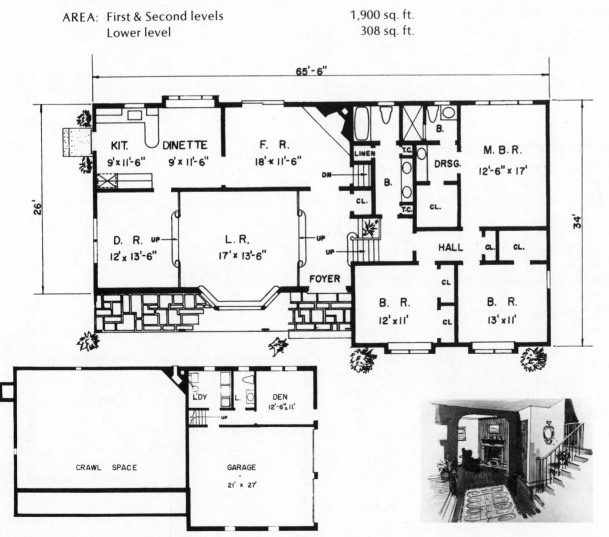

The Farley

In split-level homes, as in all other types, there are the needs of the large family to consider. Here is a plan comprising four very large bedrooms, many closets and two full bathrooms. There is also, in addition to the regular living room, dining room and kitchen, a wonderfully large porch, and just off the entrance hall and down a few steps are situated a beautiful recreation room, laundry area and lavatory. Ample space throughout is provided here for the many varied activities of the large family.

AREA: First & Second levels   1,920 sq. ft.
         Lower level   406 sq. ft.

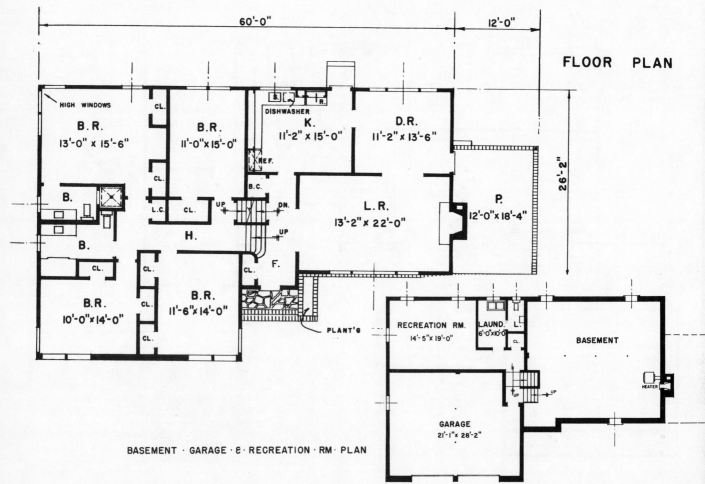

122

The Dorsey

From the circular entrance foyer, with its sweeping stairway, to the roomy secluded family room, this split-level house represents a different, modern style of home design.

Other features to be found are full-size glass sliding doors opening out on the terrace from the living, dining, and family rooms. The lower level contains a spacious two-car garage and a recreation room we feel sure you will desire.

AREA: First & Second levels   1,920 sq. ft.
        Lower level   308 sq. ft.

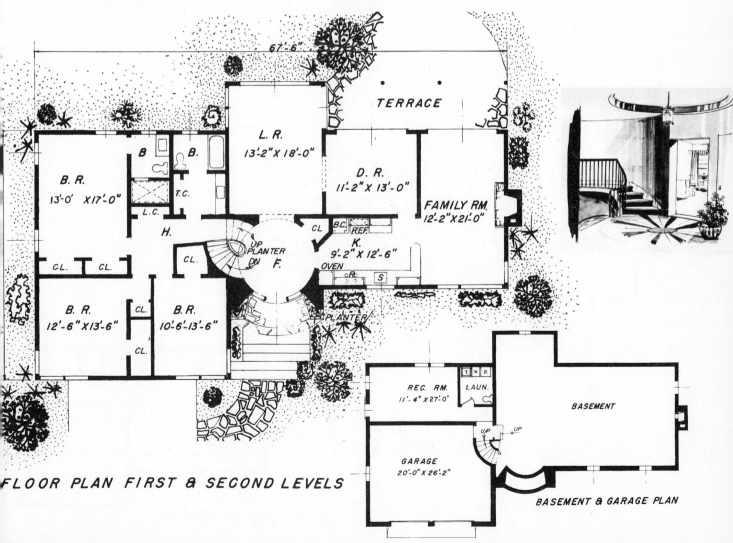

FLOOR PLAN FIRST & SECOND LEVELS

BASEMENT & GARAGE PLAN

The Delray

Who could blame you if you had your housewarming before the walls were up? Just a look at the plan tells the news that there is something wonderful about this split-level home. There's dignity and comfort in the look outside, with fieldstone and shingle set off by the handsome chimney and latticed overhang at the front door. Inside, a gracious hall, with a planting box to shield the living room, faces a curving half-stairway to the bedroom level, where privacy, spaciousness and comfort are starred, and super closets and two bathrooms (one, parent-private) are luxury extras. On the living level, fold back the screen for an entertaining ell that opens to almost 27' at its width, and screened, adds a T.V. room or study. Look forward to comfortable formality in the dining room, and for service with a smile, the kitchen-dining area at the back has just about everything to make it a model. On the basement level, a two-car garage, windowed recreation room and full basement facilities make for a total living score of "Perfect."

AREA: First & Second levels   1,924 sq. ft.
        Lower level   305 sq. ft.

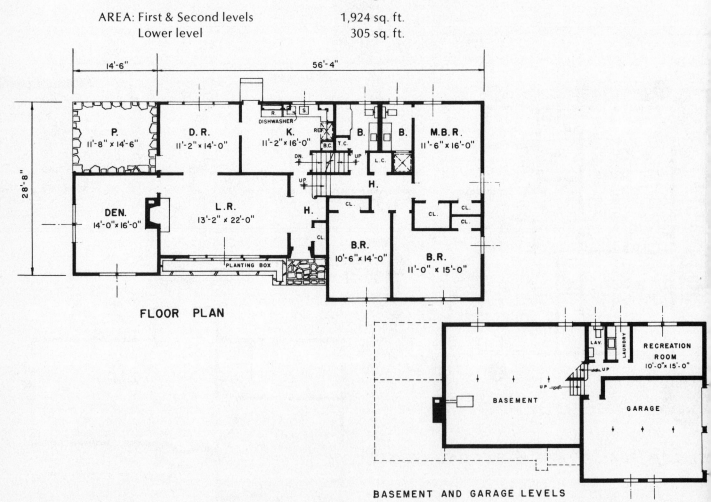

FLOOR PLAN

BASEMENT AND GARAGE LEVELS

The Monterey I

Two authentic exteriors go with this split level floor plan. The choice is yours:

A super abundance of closet space is featured in this attractive home. A very satisfying compromise between the two story and the ranch style is accomplished through the split-level. Privacy is achieved for the bedrooms off the ground level, yet they are only a half flight away from general living areas. The convenience of a "built-in" garage should not be overlooked here. Note, too, that all areas are conveniently accessible with a minimum of effort. The English Tudor has much to recommend it.

You will not go wrong in selecting this plan. In fact, when you order the blueprints, you will receive both optional elevations; so you need not decide which front elevation is your favorite right now.

AREA: Living Area   1,645 sq. ft.
        Lower level living area   280 sq. ft.
        Garage   500 sq. ft.

FIRST FLOOR PLAN

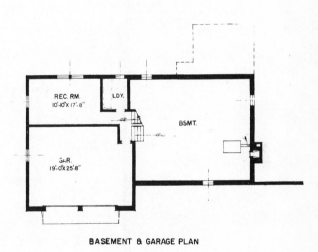

BASEMENT & GARAGE PLAN

The Monterey II

The Yardley

The exterior of this four bedroom split level design is contemporary in feeling with a pleasing combination of handsplit red cedar shingles, brick veneer, multi-unit single paned windows and trimmed with low hipped and gabled asphalt shingled roofs.

For those who can orient themselves to "split-level living" this carefully planned design, with a long list of features to recommend it, offers visual satisfaction on the outside, and practical living on the inside.

AREA: Foyer & Family room level   560 sq. ft.
       Living Area level   815 sq. ft.
       Bedroom level   1,139 sq. ft.

60'-0"

28'-6"

37'-0"

DINING RM.
12'-0"X12'-4"

KIT.
12'-2"X17'-6"

DIN.

FAMILY RM.
12'-2"X 20'-0"

L'DRY.

DW. S.

REF.

UP

UP

UP

DN

CL.

D

W

T

LIVING RM.
15'-4" X 22'-6"

UP

UP

FOYER
SLATE FLOOR

GARAGE
22'-6"X 23-8"

CL.

LAV.

CL.

33'-9"

33'-9"

CL.

BED RM.
10'-0"X13'-2"

BED RM.
12'-0"X12'-6"

DESK

CL.

CL.

DN

HALL

L.C.

CL.

CL.

BED RM.
10'-0"X14'-3"

DRESSING

BED RM.
12'-0"X17'-7"

T.C.

CL.

The Corry

This split level design, with its barn shake and board and batten siding combined with a box bay window and shutters, gives this home a truly authentic colonial appearance.

This nine room, two and one-half bath design had the housewife in mind right from the planning stage. The farm kitchen with the sink, oven, range, and refrigerator all in one unbroken "L" shape counter top provides ample table space in front of a window wall completely out of the line of traffic.

The bedroom level, which contains four rooms and two baths, has ample closet space including a walk-in type in the master bedroom. Wall space for the placement of furniture is well provided for in this design.

The laundry, only a few steps from the kitchen, is so located to allow this area to serve a dual function. Children can use this area for clean-up purposes after playing in lieu of carrying their unsightly dirt and grime through the house.

Note the king size recreation room with its front bay window and rear sliding glass door, which leads to the garden and patio, an ideal area for casual living. Connecting directly to this room is last but not least, an oversized two car garage, a must in many of today's larger families. The doors may have an alternate front or rear location depending on your lot conditions.

AREA: Entrance level   564 sq. ft.
           Living & Sleeping level   1393 sq. ft.

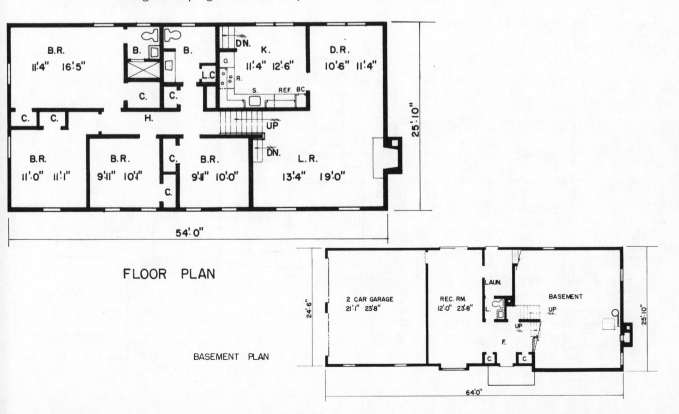

FLOOR PLAN

BASEMENT PLAN

The Roxbury

The rich graceful lines of this home extend from the overall appearance, up through the entrance, into the living room and spread throughout the whole interior. Notice the enchanting brick planting bed separating the foyer from the living room and then the graceful curving five steps leading up to the bedroom wing which consists of three very large bedrooms, spacious closet area and two full bathrooms.

AREA: First & Second levels   2,206 sq. ft.
        Lower level   960 sq. ft.

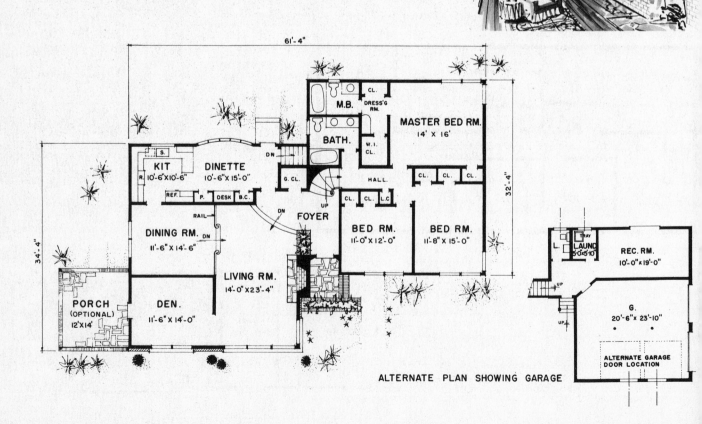

ALTERNATE PLAN SHOWING GARAGE

128

The Amity

This split-level contemporary three bedroom design combines the two-story and one-story construction features to offer the best of both; traditional separation of the living, sleeping and recreation areas, and the modern simplicity of a one floor home plan.

The gentle style of this design gives it the flexibility to fit into the suburbs or city, in the mountains or desert—just about anywhere in the country, it is a home for all seasons.

AREA: Living & Sleeping levels 2,015 sq. ft.
        Recreation & Garage   1,030 sq. ft.
        Basement   830 sq. ft.

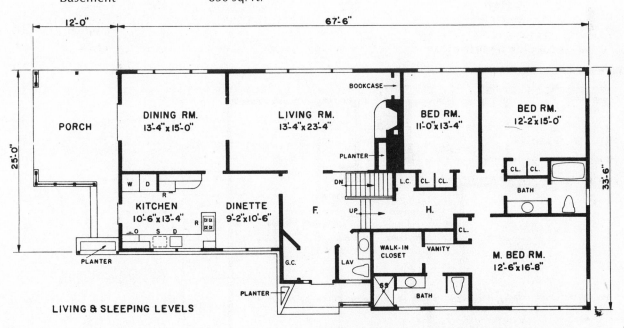

LIVING & SLEEPING LEVELS

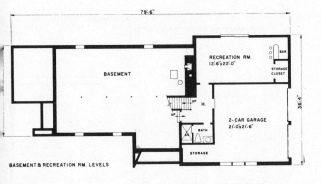

BASEMENT & RECREATION RM. LEVELS

The Ashley

A perfect study in comfort and luxury is apparent in a single glance at this design.

The large kitchen, a thru-hall entrance, and secluded bedrooms are only a few of the advantages to be found skillfully worked into this popular home plan.

AREA: First & Second levels   1,740 sq. ft.
       Third level   365 sq. ft.
       Lower level   312 sq. ft.

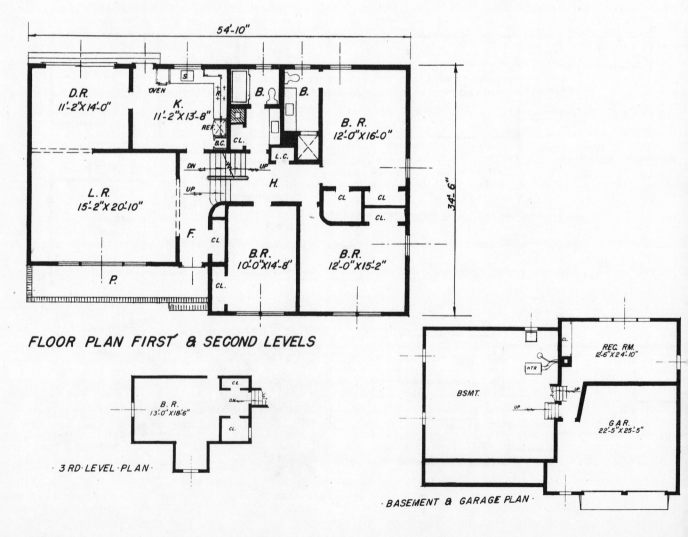

FLOOR PLAN FIRST & SECOND LEVELS

3RD LEVEL PLAN

BASEMENT & GARAGE PLAN

130

The Shrewsbury

True to its English Tudor heritage, this graceful three bedroom split level design offers great visual variety, and makes use of traditional materials such as stucco, stone, brick veneer, rough timbers and a textured shingle roof. A most impressive feature of this design is the circular two-story stone-veneer tower with its heavy oak entrance door and the circular entrance foyer which features a large open winding wrought-iron stairway leading to the upper bedroom hall and down to the recreation room and garage.

AREA: First & Second levels   2,050 sq. ft.
       Lower level   400 sq. ft.
       Garage   570 sq. ft.
       Basement   1,050 sq. ft.

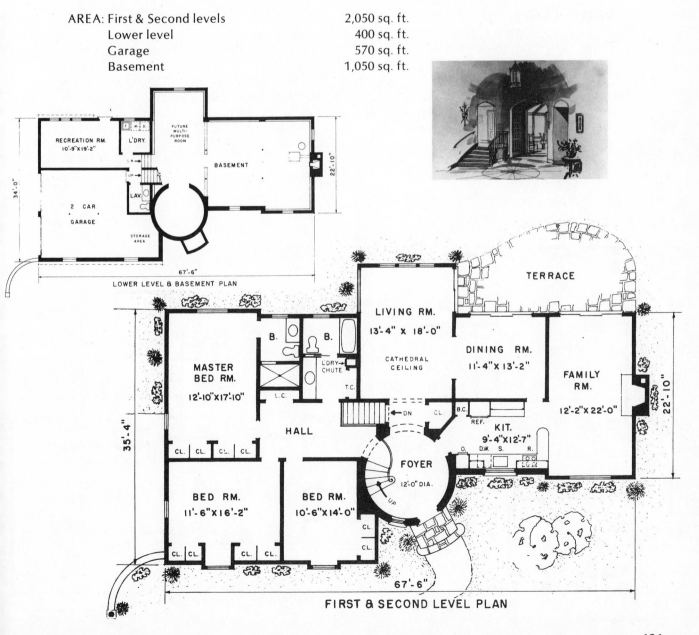

LOWER LEVEL & BASEMENT PLAN

FIRST & SECOND LEVEL PLAN

The Murray

From its two car garage at one end to the bedroom levels at the other, this split-level offers something a little different in home designing.

The handsome foyer has two large closets and a handy powder room. A long planting box separating the large dining and super-sized activity rooms adds a refreshing touch we are sure you will applaud. The kitchen and living room, yes, everywhere nothing has been spared to bring you a home that is as modern as tomorrow.

AREA: First & Second levels   2,100 sq. ft.
        Lower level   663 sq. ft.

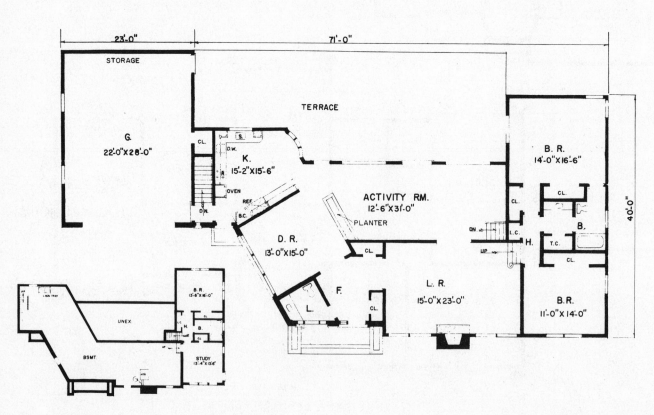

BASEMENT PLAN

The Seville

This authentic Spanish split-level design embodies the easy informality of movement indoors and exterior styling of rough stucco, circular headed windows, turned wood posts and projecting stained wood beams.

Arranged for present day living, a daylight cathedral-ceiling entrance foyer is featured with a split stair leading to the lower active area. For those who prefer a three bedroom design with an appealing combination of modern principles of privacy and split-level living with the architecture of Spanish Colonial times, here is a comfortable, livable and inviting design.

AREA: Living & Sleeping level .... 2,175 sq. ft.
    Lower level .... 966 sq. ft.
    Basement .... 1,160 sq. ft.
    Garage .... 530 sq. ft.

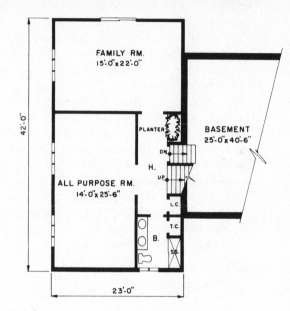

133

The Brinckley

Entrance portico, narrow clapboards, hand-split shingles and small paned windows are carefully blended to bring you this attractive split level design.

A spacious foyer with adjacent lavatory forms a most impressive reception area which leads to an oversized living room with an early American fireplace and walnut paneled wall.

The kitchen with its picturesque picture window overlooking the rear garden is only one blessing found in this design.

The glass sliding doors of the recreation room take full advantage of view, light and access to the rear yard, patio and garden areas.

Located between the kitchen and family room is a well designed laundry including closet and wall cabinets for utility storage. This room also serves as a rear service entrance.

Also, this plan provides a den with its own closet directly off the foyer, which can also be used as an overnight guest room.

There are ample sleeping accommodations with three spacious bedrooms, each having sufficient closet space. The closets in the master bedroom are out of the room in a bath with vanity and stall shower.

The oversized garage with incorporated storage area completes this split level design.

AREA: Living level   1,445 sq. ft.
        Bedroom level   805 sq. ft.

72'-0"

9'-2"

24'-4"

THIRD LEVEL

26'-4"

D'ETTE
9'-1" x 10'-10"

F. R.
13'-4" x 19'-0"

LAUN.

CL. B.C.

K.
10'-1" x 10-10

DIN.
12'-2" x 13'-4"

DN.

CL.

STUDY
11'-10" x 13'-0"

F
9'-0" x 9'-5"

DN. UP

DN.

L. R.
13'-0" x 19'-4"

GAR.
21'-8" x 23'-6"

LAV.

CL

30'-8"

CLOS

M.B.R.
12'-5" x 15'-6"

B.

CLOS

LIN.CL.

B.

H.

DOWN

B.R.
10'-6" x 14'-0"

CLOS.

B.R.
10'-6" x 12'

CLOS

22'-6"

49'-6"

FIRST • SECOND LEVEL

The Ormsby

Visions of royal living and spaciousness nestling in a countryside come quickly to mind in looking at this traditional French Provincial three bedroom split-level design.

The angular diamond glazed living room bay window, wrought iron balconied dormer windows, shutters and brick exterior with brick quoins on the end of the building, help convey a feeling of old fashioned quality.

Radiating an image of living elegance, this design exhibits artistic lines on the exterior and a lavish interior layout.

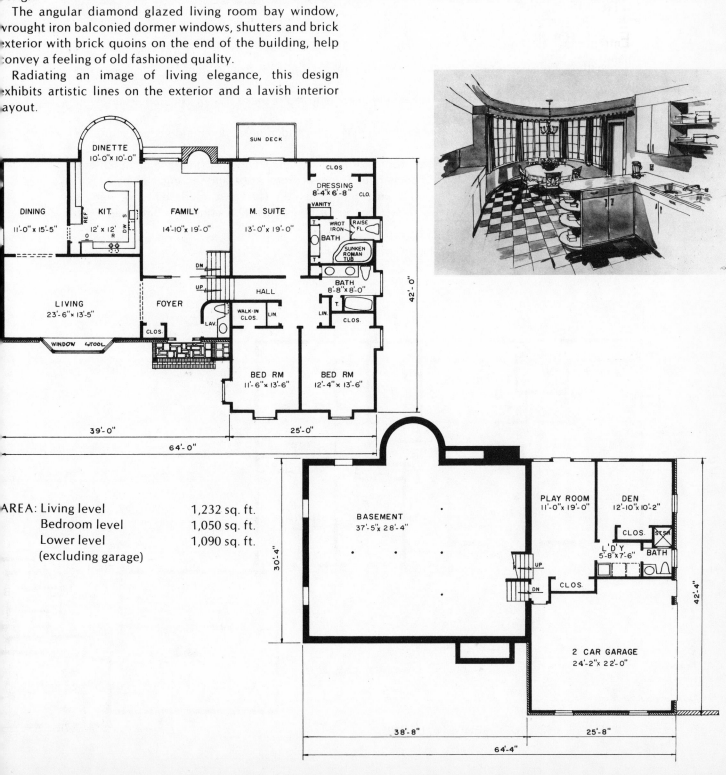

DINETTE
10'-0" x 10'-0"

SUN DECK

DINING
11'-0" x 15'-5"

KIT.
12' x 12'

FAMILY
14'-10" x 19'-0"

M. SUITE
13'-0" x 19'-0"

CLOS.

DRESSING
8'-4" x 6'-8"

CLO.

VANITY

WROT IRON   RAISE FL.

BATH

SUNKEN ROMAN TUB

DN

UP

HALL

LIVING
23'-6" x 13'-5"

FOYER

WALK-IN CLOS.

LIN.

BATH
8'-8" x 8'-0"

LIN.

CLOS.

CLOS.

LAV.

WINDOW STOOL

BED RM
11'-6" x 13'-6"

BED RM
12'-4" x 13'-6"

39'-0"   25'-0"

64'-0"

42'-0"

AREA: Living level   1,232 sq. ft.
       Bedroom level   1,050 sq. ft.
       Lower level   1,090 sq. ft.
       (excluding garage)

BASEMENT
37'-5" x 28'-4"

PLAY ROOM
11'-0" x 19'-0"

DEN
12'-10" x 10'-2"

CLOS.

STSH.

L'D'Y.
5'-8" x 7'-6"

BATH

UP

DN

CLOS.

2 CAR GARAGE
24'-2" x 22'-0"

30'-4"

42'-4"

38'-8"   25'-8"

64'-4"

The Crowley

Presenting a regal exterior of brick veneer, clapboards and picturesque shutters, this design has won widespread acclaim.

Inside we find that feature after feature has been included to make this an excellent example of experienced designing.

The lower level contains a two car garage, a play room and, if desired, a maid's room complete with full bath.

Above, we find the welcoming living room with its sunny bow window, fireplace and planters. On this level, elaborate provision has been made for meal preparation and serving. Nearby a lavatory, laundry room, and a large den containing sliding glass doors on two sides are found. Bigness best describes the slumber level with its three large bedrooms, two complete baths, and the numerous closets that abound here.

AREA: First & Second levels   2,340 sq. ft.
        Lower level   405 sq. ft.

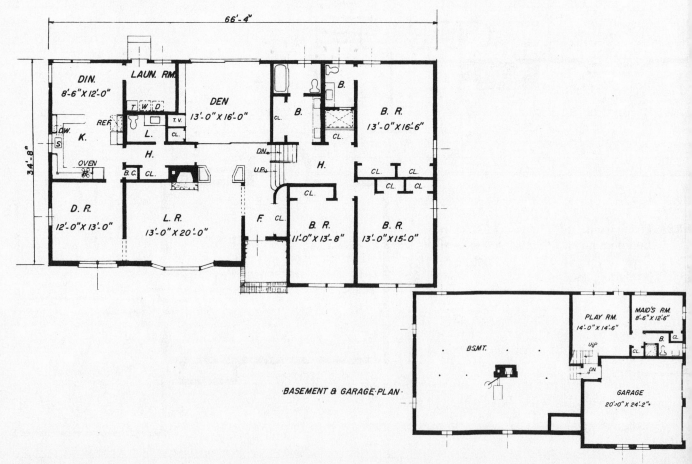

·BASEMENT & GARAGE·PLAN·

The McKinley

This multi-level plan contains many features for the large and growing family.

A ground level entrance leads from the weather protected porch to a delightful vestibule flanked by closet and powder room; then into a spacious foyer with a brick planter; folding louvered doors lead to the recreation room, which has a direct connecting stair to the kitchen—notice also the concealed laundry location off the recreation room. On the opposite side are a cozy den and fireplace well secluded for quiet moments.

A wide short stair leads up to the living area which stretches across the rear of the house and features sliding glass doors to the patio. Up again and to the front are three bedrooms and two baths—but that is not all—once more a short flight of steps and to the rear again there is space for two more bedrooms and another bath. Truly a home for the large and growing family.

AREA: Entrance level   720 sq. ft.
        Living & Bedroom levels   1,576 sq. ft.
        Fourth level bedrooms   700 sq. ft.

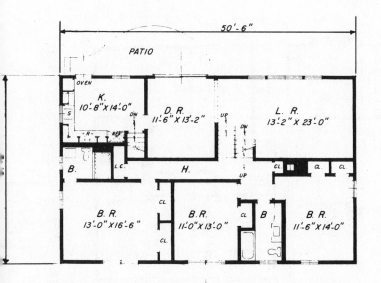

·FLOOR PLAN ·LIVING & BEDROOM LEVELS·

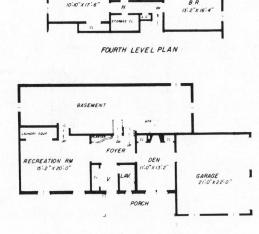

·FOURTH LEVEL PLAN·

ENTRANCE LEVEL & BASEMENT PLAN

The Thackeray

Here is the answer to your dreams.

This ten room home, complete in all respects, has all the conveniences and appeal that can be offered. Its large living room with complete wall fireplace and bay window connecting directly to an oversized dining room offers an ideal entertainment area.

The family room with front and rear exposures including a "second" fireplace opens directly to a raised terrace which overlooks the rear garden area. This area being ideal for outdoor entertaining.

Note the spacious farm type kitchen, the answer to every woman's dream, located directly off the foyer and connecting directly to the dining area and laundry room.

The bedroom level is separated from the living level by five steps and overlooks the lower foyer.

There is a large hall bath with a 6' vanity that will attract everyone's attention.

The study directly off the upper foyer can be used as a guest room.

The two car garage, including a storage area, recreation room and lavatory just eight steps down from the foyer, completes this split level design.

This home with its hand-split shingle and brick veneered exterior will surely be a "leader" in any community.

AREA: Living level   1,325 sq. ft.
        Bedroom level   1,175 sq. ft.
        Lower level   378 sq. ft.

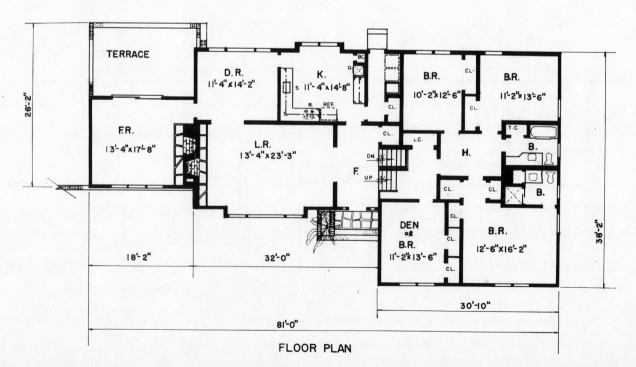

FLOOR PLAN

The Chandler

Dramatic interest and good planning prevail throughout this three bedroom contemporary split-level design. Notice how the weather protected double-door entrance is given outstanding importance by the "floor-to-ridge" glass treatment that floods the cathedral-ceiling foyer with daylight. The dinette, kitchen and family room feature a stone-faced fireplace in combination of 37 feet of open space. Two compartmentalized full bathrooms provide the ultimate in service for the bedrooms. Wide steps lead down to the sunken living room from the foyer and dining room, and directly under the bedrooms are the two-car garage, recreation room, lavatory and laundry with convenient access to the upstairs foyer.

AREA: Living & Sleeping area   2,533 sq. ft.
           Garage   552 sq. ft.
           Basement   1,981 sq. ft.

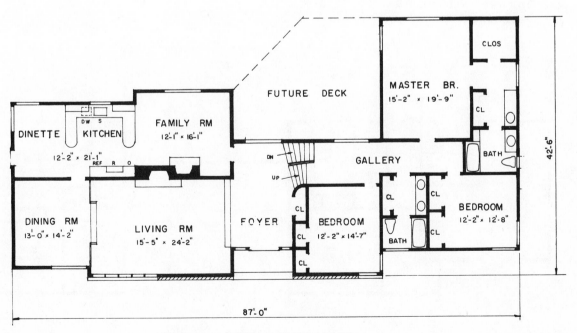

UPPER & LOWER LEVEL PLAN

The Colony

This delightfully liveable multi-level home has many unusually different features.

The long wrought iron columned portico across the front lends a warm colonial touch to the exterior and provides weather protection for the main entrance.

A spacious entrance foyer forms a most impressive reception area for guests. Leading up from this area is a balconied stair to the living room. Notice the kitchen extending to the rear and forming a bright corner-windowed breakfast area overlooking your patio.

The eight rooms and three baths in this house provide ample space for a large or growing family plus separate areas for overnight guests and entertaining.

AREA: Living Room & Bedroom levels   1,492 sq. ft.
           Entrance & Recreation levels   819 sq. ft.
           (excl. garage)

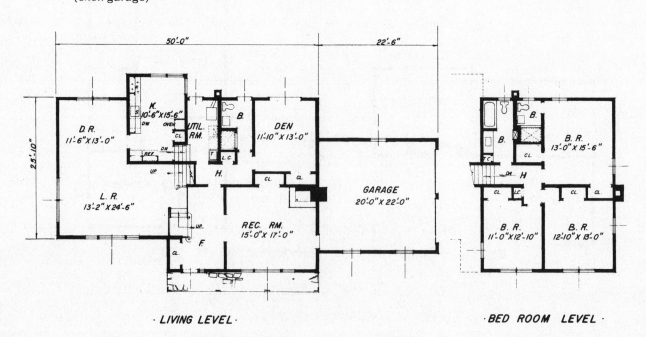

· LIVING LEVEL ·   · BED ROOM LEVEL ·

The Versailles

Luxurious master bedroom suite with glass doors leading to a private balcony . . . archway connects to dressing area . . . bath has free-form Roman tub set in tiled floor and screened-off water closet . . . two other bedrooms have double exposure . . . living room is directly to left of large foyer and adjacent to dining room . . . kitchen is enhanced by an imposing dinette area with a semi-circular floor-to-ceiling bay window . . . family room has an arched fireplace built into a brick wall . . . glass sliding wall looks over garden . . . level beneath has two-car garage, playroom leading to backyard and den . . .

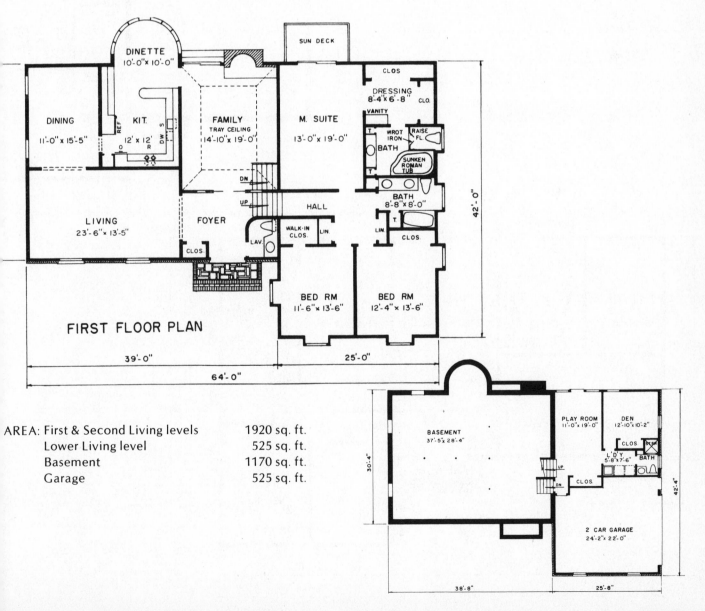

FIRST FLOOR PLAN

AREA:	First & Second Living levels	1920 sq. ft.
	Lower Living level	525 sq. ft.
	Basement	1170 sq. ft.
	Garage	525 sq. ft.

The Tamarind

Contemporary in spirit and styling is this multi-level design. It features vertical red-wood siding, random width fieldstone, low-pitched sweeping rooms and generous fenestration by the use of clerestory and canopied casement windows to create architectural interest. The three-level scheme for living makes it ideal for today's active life style.

AREA: First & Second levels   2,290 sq. ft.
    Lower level   550 sq. ft.
    Garage   540 sq. ft.
    Basement   1,180 sq. ft.

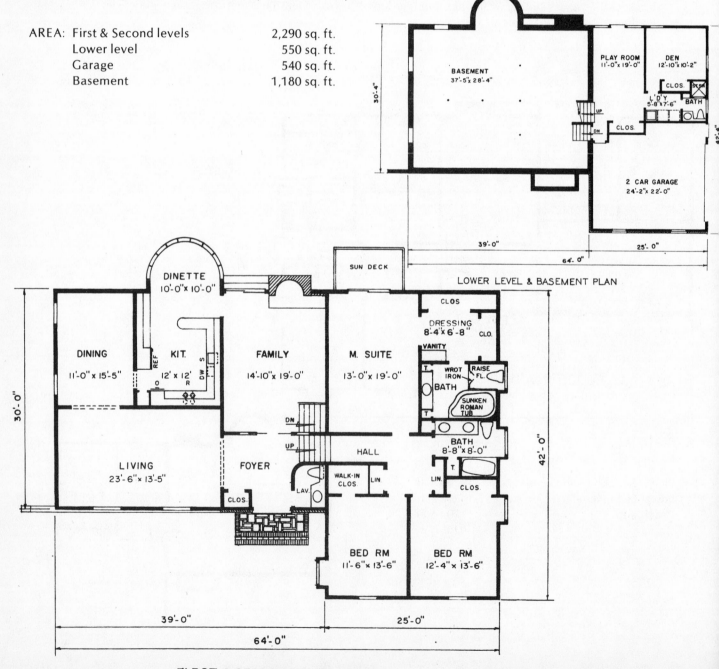

LOWER LEVEL & BASEMENT PLAN

FIRST & SECOND LEVEL PLAN

MULTI-LEVEL

Sometimes called by other names, such as Hi-ranch and Bi-level. In this type of house, the front foyer is at ground level, with a stairway upward to the main living area and another downward to what would ordinarily be the basement. Because the basement is raised out of the ground enough to permit windows above ground, the area is utilized for living purposes and usually contains a recreation or informal room.

The Middlebury

The perfect home for the narrow lot. This plan can fit on most 50' lots.

The upper level consists of six rooms and bath totaling 965 square feet. The plan is ideal for a young budget minded family.

Containing three ample size bedrooms with individual closets, provide adequate sleeping accommodations.

The kitchen directly off the upper foyer is the center of circulation for this compact plan. The balconied living room which adjoins the dining room creates a glamorous effect of the combined area.

The Adams

Modern living is the idea of this bi-level. Picture yourself in this house, the envy of the neighborhood, the satisfaction of being the pace setter. Clean design and a friendly entrance invite you to enjoy the roominess of three bedrooms, two baths, large step saving kitchen and informal living room. The lower level continues the mood, opening to a creative recreation room with corner fireplace and window door access to the patio. Note the privacy yet central location of the den. Inside and out this house is the paragon of modern thought and execution.

AREA: Upper level   1,188 sq. ft.
       Lower level   687 sq. ft.

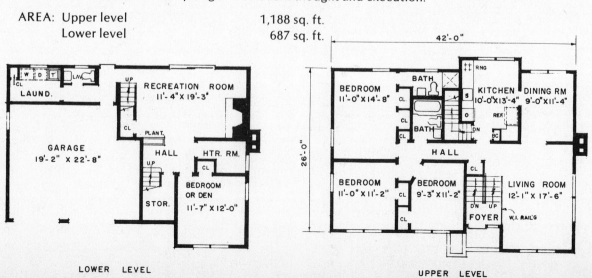

LOWER LEVEL   UPPER LEVEL

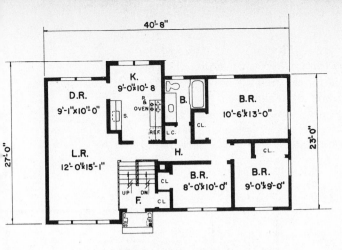

FIRST FLOOR PLAN

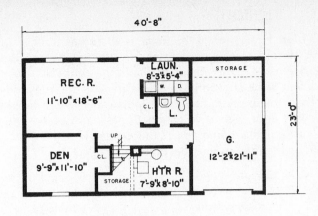

BASEMENT PLAN

The lower level with its recreation room, den, laundry and oversized garage complete this truly fine plan.

The exterior covered entrance platform, together with its clapboard and vertical siding, will be an eye catcher in any neighborhood.

AREA: Living and bedroom level   965 sq. ft.
Recreation and garage level 935 sq. ft.

The Monticello

All the charm and elegance of Colonial styling is embodied in this bi-level three bedroom ranch home with its gabled roof entrance portico, shutter-trimmed multi-paned windows, a handsome entrance and a facade of brick veneer and red cedar clapboards.

Inside, the "mid-level" or "split entry" foyer provides direction to either the upper or lower level.

With only 1,205 square feet of well utilized space on the main level and the entire house only 42 feet 6 inches wide, a large plot is not required. This house is designed for economy in construction and will provide good living for a fairly sizeable family.

AREA: Living level   1,205 sq. ft.
Lower level   541 sq. ft.
Garage   628 sq. ft.
Deck   83 sq. ft.

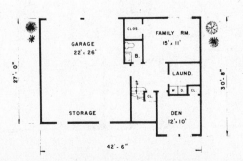

LOWER LEVEL PLAN

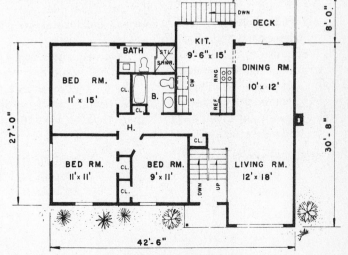

UPPER LEVEL PLAN

The Kingston

The basic living area of this home is all on one floor—"ranch style." The unusual feature of the "split" entrance vestibule creates an exterior appearance of the formal "two story colonial." This entrance feature also makes available the entire basement area for finished liveable rooms. The basement being only partially below grade in the front and on grade in the rear.

Ranch homes are normally the most expensive to build per square foot of living area, but this ingenious arrangement of entrance making available the finished living areas in the basement tends to put this home nearer the category of the 2 story home for economy of construction based on the total square feet of living area.

AREA: First floor   1,650 sq. ft.
        Basement, Recreation room,
           Den & Lavatory   600 sq. ft.

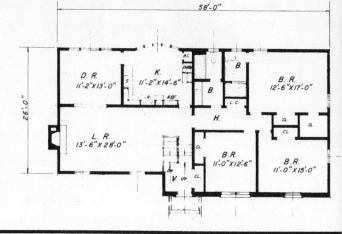

BASEMENT & GARAGE PLAN

The Gateway

All the charm and elegance of the South are embodied in this traditional Southern Colonial three-bedroom bi-level design with its two-story central portico, massive square pillars, and combination of brick veneer and natural wood red shingle cedar exterior.

Because of the economics of space utilization and construction, increasing numbers of today's home buyers are finding the *raised ranch* to their liking for comfortable living, where routine family activity centers on the upper level, while the lower level is a spacious asset for entertaining and relaxation.

AREA: First floor   1,430 sq. ft.
        Basement   880 sq. ft.
        Garage   550 sq. ft.

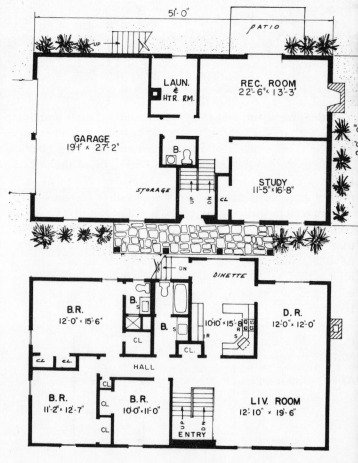

The Vanderbilt

Increasing numbers of today's new home owners are finding the "raised ranch" to their liking because of the economics of construction and space utilization.

This design, which retains all the romantic charm of the Old English Tudor, is a typical raised ranch plan with a complete three-bedroom one-floor living unit set on top of a daylight basement that offers extra space that does not look subterranean. The traditional styling of the exterior is enhanced by the wood shingles, stucco, boxed living room bay, hand-hewn timber and small paned windows.

The solid enduring look of this bi-level design, with value as its prime ingredient, will be a source of pride in any neighborhood.

AREA: Upper level   1,458 sq. ft.
        Lower level, Living area   956 sq. ft.
        Garage   462 sq. ft.

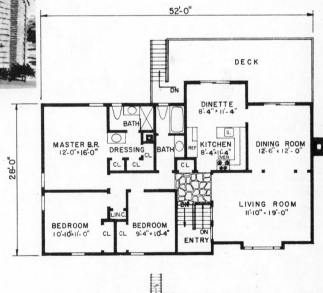

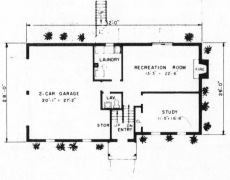

147

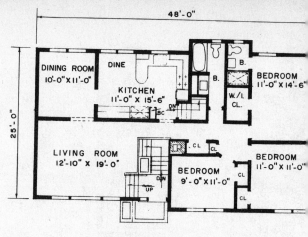

UPPER LEVEL

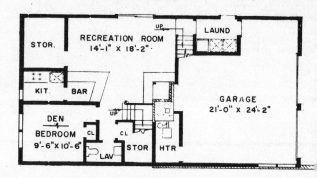

LOWER LEVEL

The Carlyle

Three large bedrooms and double acting bath-dressing area for the master chamber provides for the privacy and roominess desired. The large kitchen and dinette with adjacent dining room complement the airy living room highlighted by its stone planter.

Downstairs, the recreation room, corner fireplace, bar and kitchenette provide the excellent opportunity for entertaining and just family fun. The spare room with lavatory is perfect as a den or for overnight guests. Notice the kitchen stair for easy access to laundry, garage and recreation room.

Extras throughout add up to exciting living in this roomy little house.

AREA: Upper level   1,266 sq. ft.
Lower level   400 sq. ft.

The Pendrey

Modern living is the idea of this bi-level. Picture yourself in this house, the envy of the neighborhood, the satisfaction of being the pace setter. Clean design and a friendly entrance invite you to enjoy the roominess of three bedrooms, two baths, step saving kitchen with large dinette, cathedral ceiling living room and dining room. The lower level continues the mood, opening to a creative recreation room with fireplace and sliding glass door access to patio. Note the privacy yet central location of the study. Inside and out this house is the paragon of modern thought and execution.

AREA: Upper level   1,470 sq. ft.
Lower level   722 sq. ft.

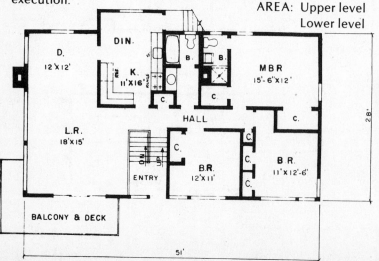

FIRST FLOOR

BASEMENT

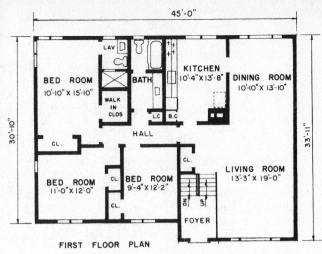

FIRST FLOOR PLAN

45'-0"

30'-10"

33'-11"

BED ROOM 10'-10" X 15'-10"
LAV
BATH
KITCHEN 10'-4" X 13'-8"
DINING ROOM 10'-10" X 13'-10"
WALK IN CLOS
LC. B.C.
HALL
CL.
BED ROOM 11'-0" X 12'-0"
CL. BED ROOM 9'-4" X 12'-2"
CL.
LIVING ROOM 13'-3" X 19'-0"
CL.
DN UP
FOYER

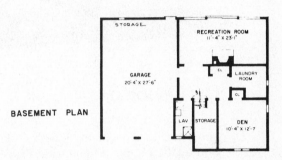

BASEMENT PLAN

STORAGE
RECREATION ROOM 11'-4" X 23'-1"
GARAGE 20'-4" X 27'-6"
CL. LAUNDRY ROOM
CL.
LAV STORAGE
DEN 10'-4" X 12'-7"

The Copeland

All the features of larger more expensive homes are incorporated in this moderate sized bi-level home.

Three full bedrooms and two baths on the sleeping level—notice the hall bath is split for more efficient use with privacy.

A spacious kitchen with large bright and open space for eating at the windows plus a full size dining room.

At the lower level there is a tremendous recreation room, separate den (could be used as a 4th bedroom), a full bath, ample storage and an oversize two-car garage.

AREA: Upper level   1,446 sq. ft.
           Lower level   713 sq. ft.

The Webster

Refreshingly different is this split entrance three bedroom colonial design.

The exterior is underlined by wrought iron, brick and clapboards.

A cheerful corner fireplace completes the spacious recreation room. Provision has also been made for a maid's room and lavatory in this bi-level design.

AREA: Upper level   1,555 sq. ft.
           Lower level   750 sq. ft.

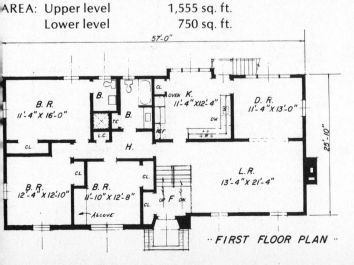

57'-0"

25'-10"

B. R. 11'-4" X 16'-0"
OVEN K. 11'-4" X 12'-4"
D. R. 11'-4" X 13'-0"
REF
H.
B. R. 12'-4" X 12'-10"
B. R. 11'-10" X 12'-8"
L. R. 13'-4" X 21'-4"
UP F DN
ALCOVE

FIRST FLOOR PLAN

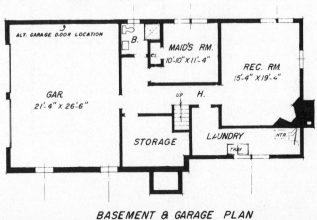

ALT. GARAGE DOOR LOCATION
B.
MAID'S RM. 10'-10" X 11'-4"
GAR. 21'-4" X 26'-6"
REC. RM. 15'-4" X 19'-4"
UP H.
STORAGE
LAUNDRY
HTR.
TRAY

BASEMENT & GARAGE PLAN

The Osborne

The intricate three-level "Raised-ranch" design of this contemporary plan produces unusually good traffic pattern. The entrance foyer acts as a distribution point—up to the living and sleeping areas by means of a six riser circular wrought iron stairway and down seven risers to the lower level that features a wood-paneled recreation room with fireplace, a maid's room or den with stall-shower bath and a two car oversize garage. Vertical lines and window treatments are emphasized on the dramatic exterior and enhanced by the floating wood deck that provides outdoor living and dining off the kitchen-dinette. The three bedrooms are well and conveniently serviced by the two baths and are supplied with ample closet space.

AREA: Foyer and Upper level   1,978 sq. ft.
         Lower level   1,358 sq. ft.
         Garage   620 sq. ft.

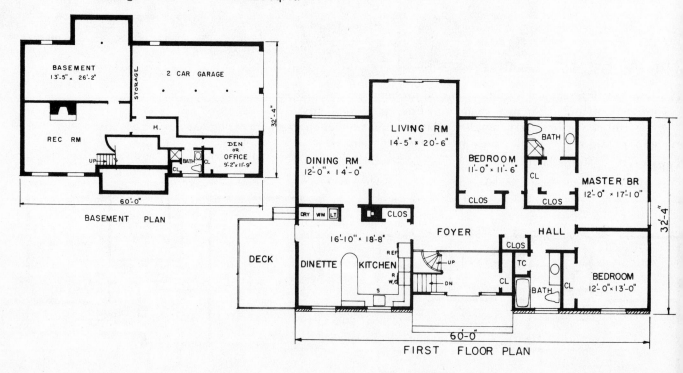

TWO STORY

Cost, on the basis of amount per square foot, is usually lower than other types of houses. Bedrooms have more privacy. Should have at least two bathrooms. More rooms can be built on less land. Many different architectural styles are available.

The Town-House

EXTERIOR— Authentic Early American salt-box; white clapboard accented by louvered shutters, huge capped chimney, interesting treatment of garage with front gambrel roof and smart entrance.

FIRST FLOOR— Spacious central entrance hall through to kitchen, coat closet, broom closet, attractive stair well, living room 20' long, fireplace, wood storage; dining room opens onto covered porch for outdoor eating, built-in china closet; kitchen efficient "U" type with extra cabinets on 4th wall; snack bar; downstairs lavatory; separate laundry; direct covered access from garage to house.

SECOND FLOOR— Three bedrooms and bath; provision for 2nd bath and future study or additional bedroom; abundance of closets; extra storage space under eaves in main bedroom. Real attic over second story for dry storage of family relics. It is reached by stair from small bedroom.

AREA: First floor   704 sq. ft.
   Second floor   704 sq. ft.
   Future study & Bath   186 sq. ft.

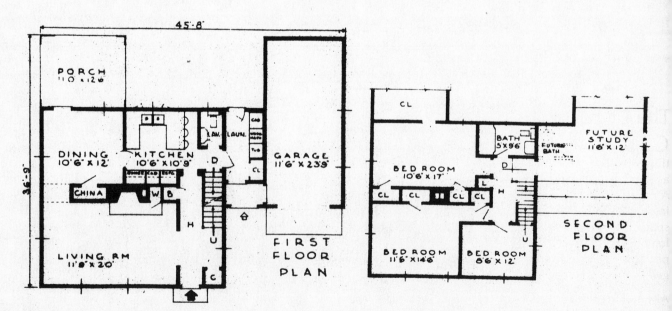

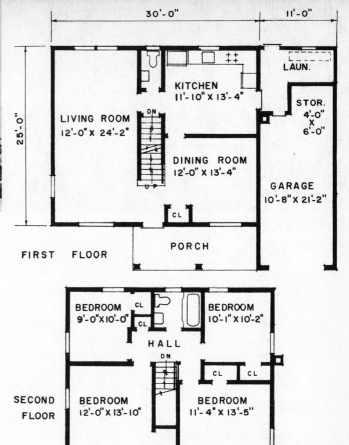

FIRST FLOOR

KITCHEN 11'-10" X 13'-4"

LAUN.

STOR. 4'-0" X 6'-0"

LIVING ROOM 12'-0" X 24'-2"

DN

UP

DINING ROOM 12'-0" X 13'-4"

GARAGE 10'-8" X 21'-2"

CL

PORCH

30'-0"   11'-0"

25'-0"

SECOND FLOOR

BEDROOM 9'-0" X 10'-0"

CL

BEDROOM 10'-1" X 10'-2"

HALL

DN

BEDROOM 12'-0" X 13'-10"

BEDROOM 11'-4" X 13'-5"

CL

CL

CL

CL

The Glen-Dale

Typically colonial, this house means home to many families whose ideas of comfort are as modern as tomorrow. Conveniently grouped around the central stair this house offers a large living room windowed at each end, efficient kitchen large enough for informal family dining and dining room for formal occasions. The second floor contains 4 bedrooms and bath with more than ample closets. Coupled with attached garage and entrance porch this house is an asset to any community.

AREA: First floor   810 sq. ft.
         Second floor   750 sq. ft.

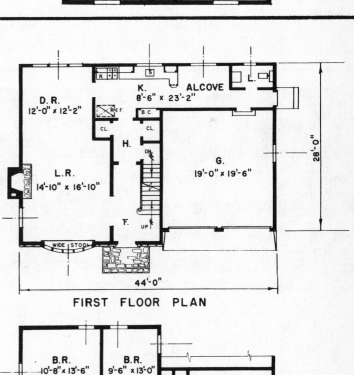

D.R. 12'-0" x 12'-2"

K. 8'-6" x 23'-2"

ALCOVE

CL.

B.C.

CL.

REF.

H.

L.R. 14'-10" x 16'-10"

DN

G. 19'-0" x 19'-6"

F.

UP

WIDE STOOP

28'-0"

44'-0"

FIRST FLOOR PLAN

B.R. 10'-8" x 13'-6"

B.R. 9'-6" x 13'-0"

CL.

CL.

B.

H.

B.R. 13'-6" x 14'-4"

DN

CL.

B.R. 12'-0" x 17'-0"

L.C.

B.

CL.

SECOND FLOOR PLAN

153

The Glen-Brook

Convenience is added to the valued privacy afforded in this two-story plan by incorporating a built-in two-car garage and lavatory directly adjacent to the service entrance. A modern touch is given to the interior by opening the dining room to the living room, thereby creating the effect of spaciousness from front to rear with an unbroken area 20' long. The additions of a room over the garage provides ample space for a child's playroom or an extra room for guests—a total of four bedrooms in all.

AREA: First floor   885 sq. ft.
         Second floor   1,050 sq. ft.

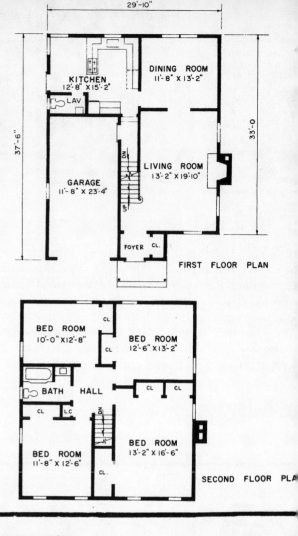

FIRST FLOOR PLAN

BED ROOM 10'-0"X12'-8"

BED ROOM 12'-6"X13'-2"

BATH HALL

BED ROOM 11'-8" X 12'-6"

BED ROOM 13'-2" X 16'-6"

SECOND FLOOR PLAN

The Dell-Wood

"Practical-minded" you sense upon entering the separate foyer with its ample closet.

The well planned kitchen-dinette and adjoining dining room make serving an ease.

The spacious living room with a fireplace calls for relaxation.

There are a large bath and four large bedrooms on the second floor with ample closets for all. The extra large closet in front could be converted to a private bath for the master bedroom.

This well planned home fits everyone's needs and also fits on most small lots.

AREA: First floor   899 sq. ft.
                      (excl. garage)

Second floor   990 sq. ft.

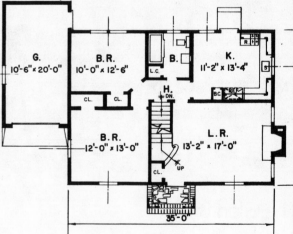

FIRST FLOOR PLAN

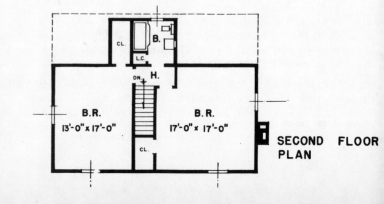

SECOND FLOOR PLAN

The Oak-Hill

Here is an economical plan developed in the homey New England salt box style. For those who love the endearing charm of days gone by we have combined all the conveniences of modern living with the appearance of tradition. Where two bedrooms are sufficient for the immediate needs of a family, this house may be built with the second floor unfinished. Later, when the need arises, the two enormous second floor rooms may be finished off for children's bedrooms or a play room.

AREA: First floor   910 sq. ft.
       Second floor   725 sq. ft.

The Green-Briar

Brick veneer, beveled siding, shutters, bow window and arch-sheltered entry and an overhanging second floor all add up to a charming exterior in this four bedroom, full basement two-story house with a modified horizontal appearance.

A dramatic double door entrance leads into an open-stair reception area. Straight ahead the all-purpose wood-paneled family room located between the kitchen and living room is made the hub of family and guest entertaining activity.

AREA: First floor   925 sq. ft.
        Second floor   864 sq. ft.

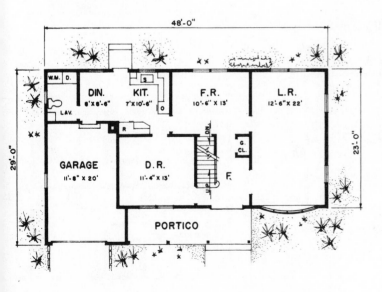

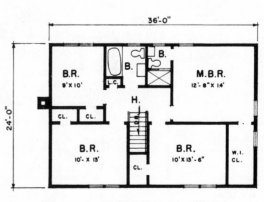

The Lynn-Brooke

Here is a charming colonial for the budget minded. Its center foyer with adjoining lavatory leads directly to all major living areas. The large family size kitchen is a large family room with front and rear exposures, including glass sliding doors opening to the rear patio. Just a few steps from the kitchen is a combination laundry-mud room opening directly to the rear yard.

The four bedrooms and two baths on the second floor with its generous closets offer ample space for a growing family.

Note the ideal out of the way storage in the front of the two car garage.

AREA: First floor   995 sq. ft.
        Second floor   837 sq. ft.
        Total   1,832 sq. ft.

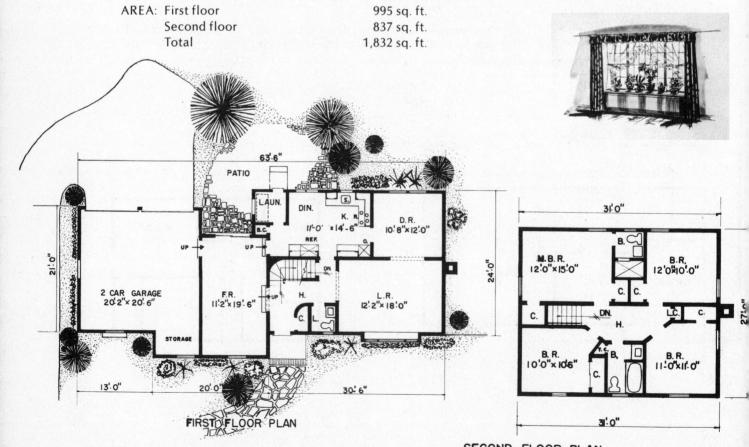

FIRST FLOOR PLAN

SECOND FLOOR PLAN

The Oak-Ridge

Typically New England Colonial, this two-story home has all the conventional characteristics of yesterday's appearance plus all the modern conveniences of today's planning and equipment.

The built-in garage and den wing also contains a convenient powder room for guests plus space above for an additional bedroom which could be left unfinished to provide for a growing family.

The main body of the house contains a modern working kitchen plus space for eating, a full sized dining room and living room all serviced by a through hall entry containing the stair. Upstairs are three good sized bedrooms and two baths; plenty of closets and of course the extra fourth bedroom as mentioned before.

AREA: First floor   950 sq. ft. excl. porch & garage
          Second floor   1,058 sq. ft.

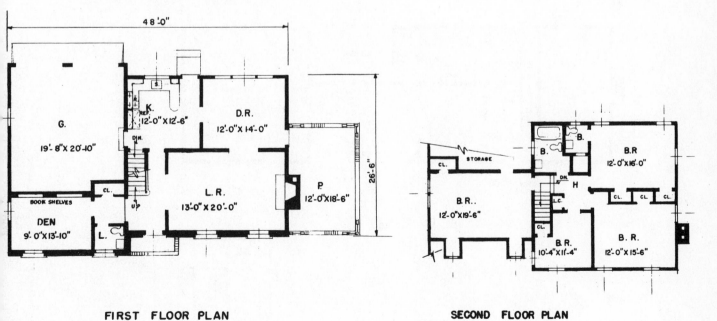

FIRST FLOOR PLAN   SECOND FLOOR PLAN

The Sher-Brooke

This seven room colonial with attached two car garage has everything that a family needs. The full length living room with its distinguished fireplace enjoys three exposures. Note the size of the kitchen-dinette, truly a "family size". Adjacent to the kitchen is the dining and family room with its sliding doors leading out to a covered porch. The first floor lavatory is readily accessible for the children from the outside.

Upstairs are three large bedrooms and two baths, plus seven large closets, high-lighted by the large walk-in closet in the master bedroom. This New England Colonial with its narrow clapboard and wood shingle exterior would be an asset to any community.

AREA: First floor   1,030 sq. ft.
         Second floor   805 sq. ft.
         Total   1,835 sq. ft.

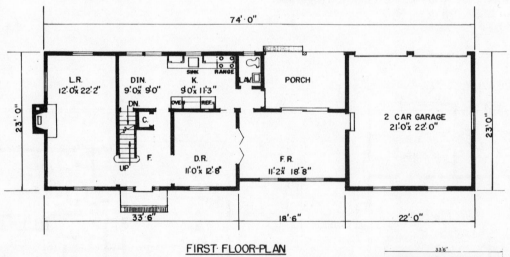

FIRST·FLOOR·PLAN

SECOND FLOOR PLAN

The Wynne-Wood

Adherents of the two story design will find much to admire in this four-bedroom house of casual contemporary styling that will set it apart from all others in the neighborhood in a strikingly lovely way. The brick planter in front of the plateglass picture window of the living room, the vertical boarding exterior, the sunroof over the entry court and the interior layout give it a totally "today" look, but its beauty is classic enough to make it endure for many years.

Guests enter through a partially covered area between the attached two car garage and the house, into a spacious foyer which leads to a sunken living room to add even more grandeur.

Upstairs, the four bedrooms complete a plan which retains all the good qualities of modern living. The master suite has cross-ventilation, a basined and mirrored vanity and a glass enclosed stall shower.

The Spartan simplicity of this contemporary, low pitched roof, two story design can provide the ultimate in modern living, at home in any surrounding, in any part of the country.

AREA: First floor   1,044 sq. ft.
        Second floor   1,040 sq. ft.
        Basement   1,044 sq. ft.
        Garage   528 sq. ft.

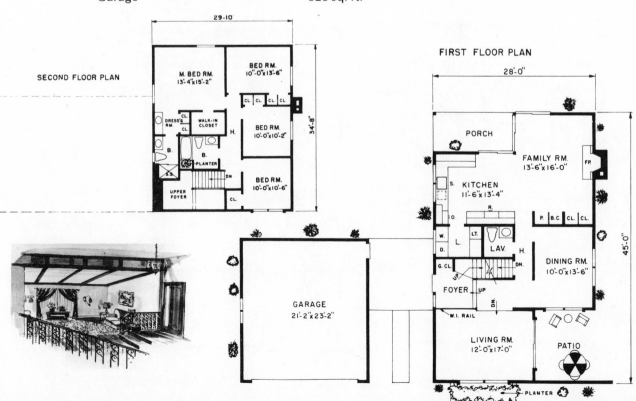

The Coventry

The charm of this Tudor adaptation, reminiscent of Old England, could hardly be improved upon. Its fine proportions and exquisite use of exterior materials of half-timber, stucco, multi-paned windows, steep hipped trimmed circular entrance result in a most distinctive home. Even the attached garage with its hipped dormer and diamond shaped leaded window, and the extended wall add impact to this design.

Designed to contribute to a feeling of personal luxury, the master bedroom suite has a dressing area with three closets and a private bath with mirrored vanity and tiled shower stall.

Each of the other three bedrooms is served by the main bath, which has a tub and a full length mirrored vanity.

AREA:
First floor	1,082 sq. ft.
Second floor	916 sq. ft.
Garage	506 sq. ft.
Patio	160 sq. ft.

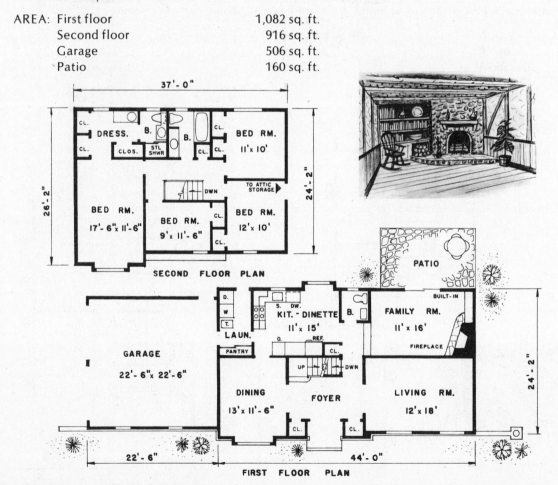

The Strathmore

The fine proportion of this impressive exterior, with its stone and brick veneer, half-timber, stucco, half-dormers, diamond and multi-paned windows, is distinctly English, and identifies the Tudor heritage of this two-story, four-bedroom design. Adding impact are the stone trimmed arched entrance, massive brick chimney and the extended decorative wall.

Visual variety, so pleasing outside, is continued indoors with a breathtaking array of highlights that will cater to the whims of a large family.

Designed to contribute to a feeling of personal luxury, the lavish secluded master bedroom suite has a dressing area with three closets and a private bath with mirrored vanity and a glass enclosed tiled shower stall.

Each of the other three bedrooms is of modest size and is served by the main bath.

AREA: First floor   1,094 sq. ft.
           Second floor   934 sq. ft.
           Garage   506 sq. ft.
           Patio   160 sq. ft.

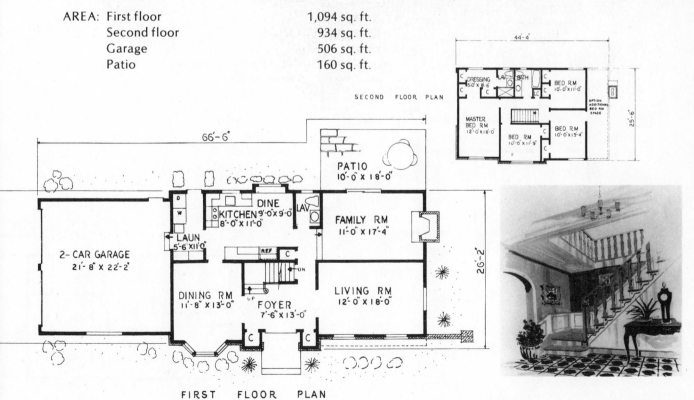

SECOND FLOOR PLAN

FIRST FLOOR PLAN

The Old-Greenwich

The gambrel roof, front facade of handsplit red cedar shakes, small paned windows, paneled entrance doors with sidelights, shedroof dormers and large chimney add to the free interpretation of this eight room two-story design.

The first floor arrangement provides a central entry hall around which all the primary living areas are oriented within an area of 1,145 square feet.

An attractive staircase leads from the entrance foyer to the four bedrooms and two baths on the second floor.

This is a good, practical and attractive house with comfort and livability built-in.

AREA: First floor   1,145 sq. ft.
        Second floor   1,145 sq. ft.

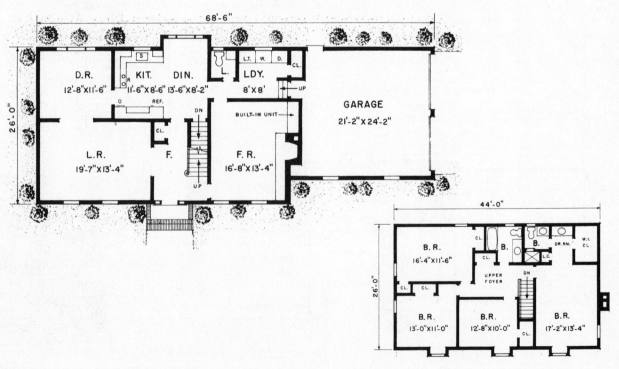

The River-Crest

Nobody has yet come up with a more economical way of housing a large family on a modest lot than with a two-story residence, and in these days of rising land prices, the financial advantage of one set of rooms atop another is greater than it ever was. Typical of many houses built during the Colonial days, this four bedroom house has the traditional small-paned, shuttered windows, beveled red cedar clapboards accented with vertical corner boards for a general air of comfort and hospitality.

AREA: First floor   1,145 sq. ft.
       Second floor   1,145 sq. ft.

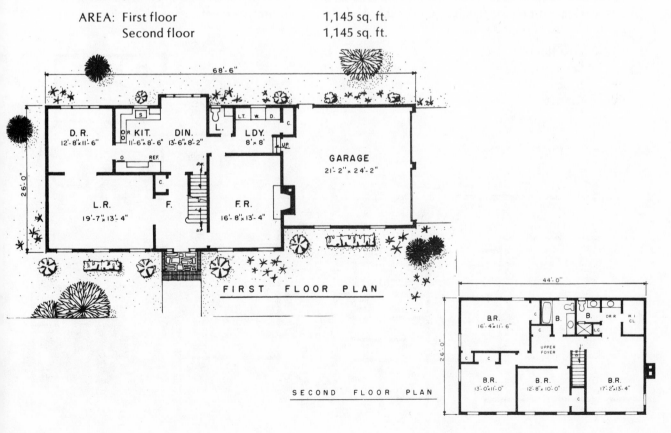

FIRST FLOOR PLAN

SECOND FLOOR PLAN

The Eton

The rediscovered charm of this Tudor design is reminiscent of the quiet dignity and flavor of early English country living. This bygone style has returned to popularity, and the reasons are many; it bespeaks solidity—its adzed timbers on stucco walls, massive brick chimney with protruding chimney pots, wavy siding, steep roofs of varying heights and diamond-paned windows are some of the basic characteristics of this "miniature-sized" version.

Although this home looks quite impressive from the outside, it is not quite as large or expensive as it looks.

The octagonal central tower that accommodates the entrance foyer is larger than some rooms; this two-story, full basement, three bedroom house will comfortably meet the needs of the average family for present day living.

AREA: First floor   1,182 sq. ft.
       Second floor   818 sq. ft.
       Basement   962 sq. ft.
       Garage   662 sq. ft.

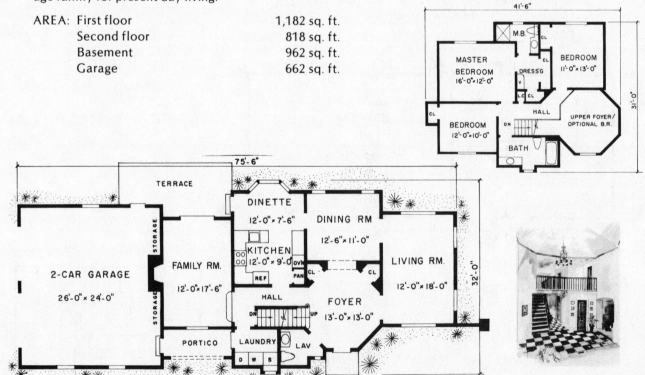

The Ferndale

The three-level entry of this staggered roofline two-story house produces unusually good traffic patterns. Distinction is added by the warm combination of hand-split shingles, clerestory windows and random-width fieldstone veneer. The entrance foyer acts as a central distribution point, from there you can go down two steps to the living room; down three steps to the family room, dining-kitchen area or walk upstairs to the four bedrooms on the second floor.

The comfort and convenience of this design is as modern as tomorrow.

AREA: First floor   1,196 sq. ft.
        Second floor   808 sq. ft.
        Garage   495 sq. ft.

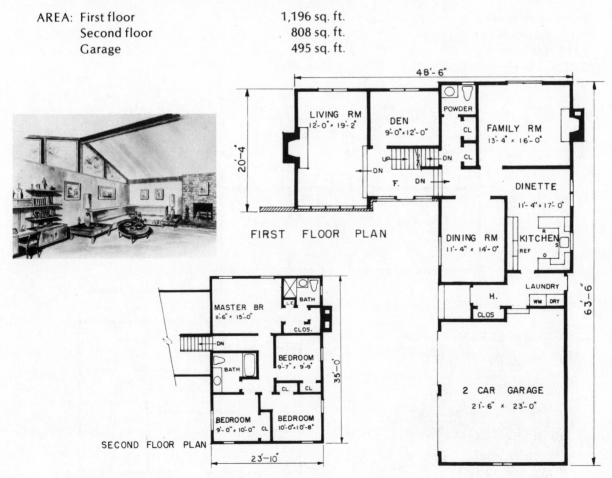

FIRST FLOOR PLAN

SECOND FLOOR PLAN

The Beaumont

When it is desired to create a good first impression in a house of moderate size, the French Provincial is often the answer. This architectural style is derived from native French architecture and has held its popularity throughout the years because it has a special kind of elegance.

The basic plan, consisting of four bedrooms and 2½ baths, is rectangular for cost-saving and has overall dimensions, including the two car garage, of 66 feet by 28 feet 6 inches.

Its combination of red cedar shingles, brick veneer, diamond-paned windows over the recessed entrance and the hipped roof will be stylish for many years to come. The sides and rear are sheathed in red cedar wood shingles. Other design elements that contribute to the attractive exterior appearance of the house are the curved brick door and window heads, the continuous dentil moulding around the eave of the roof, and the louvred cupola over the garage wing.

AREA:
First floor	1,052 sq. ft.
Laundry room	48 sq. ft.
Second floor	925 sq. ft.
Basement	1,100 sq. ft.
Garage	529 sq. ft.

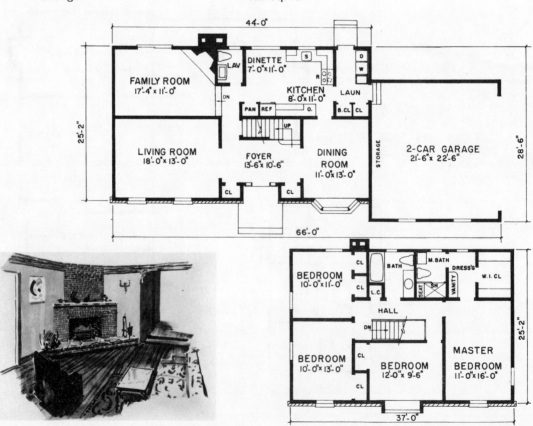

The Cedar-Wood

This stately narrow-clapboard and vertical batten traditional home is a comfortable one, built around a family that enjoys the traditional pleasures—a warm fireplace in the family room, a holiday meal in the formal dining room or a friendly party in the large living room.

This four bedroom design reflects the comfort built within, and its exterior makes it right at home in the city, suburb or country. The wood-paneled family-room with its brick fireplace has access to the paved patio, and the bay-windowed breakfast room provides a cheerful addition to the kitchen.

Inside—the unusually large and impressive foyer is the key to efficient circulation, distributing traffic effectively throughout the first floor and the second floor bedrooms.

For all-around privacy and economy of "two-story" construction, this plan is ideal for a growing family with a taste for traditional flavor.

AREA: First floor   1,102 sq. ft.
        Second floor   931 sq. ft.

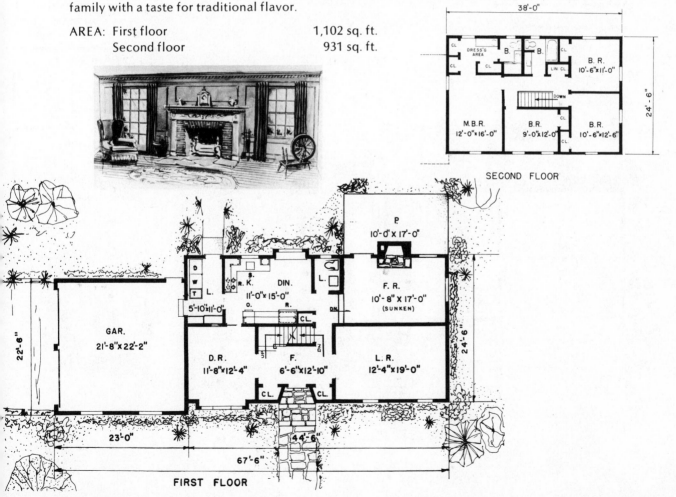

The North-Gate

Here is a charming garrison colonial for the budget minded. Its center foyer with adjoining lavatory leads directly to all major living areas. The family size kitchen with bay-windows overlooking the rear garden is every woman's delight. Adjoining the kitchen is a family room with glass sliding doors opening to the rear patio. Just a few steps from the kitchen is a combination laundry-mud room containing three individual closets.

The four bedrooms and two baths on the second floor with its generous closets offer ample space for a growing family.

AREA: First floor   1,112 sq. ft.
         Second floor   960 sq. ft.
         Total   2,072 sq. ft.

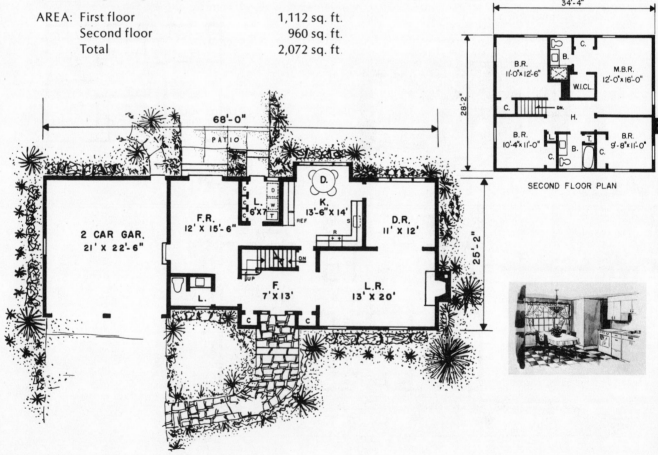

SECOND FLOOR PLAN

FIRST FLOOR PLAN

The Lowell

One of the most interesting characteristics of the popular Dutch colonial is the "Queen Anne," which features the space creating gambrel roof of two different angles, and the dormer windows in the lower of the two roof slopes.

Inside—the elevated living room, three steps above the main living level, creates a dramatic balcony view of the dining room with its decorative wrought iron railing, cathedral ceiling and lovely box-bayed window.

The three roof dormers enrich the old-fashioned enchantment of this four-bedroom Dutch Colonial design.

AREA: First floor   1,176 sq. ft.
        Second floor   896 sq. ft.

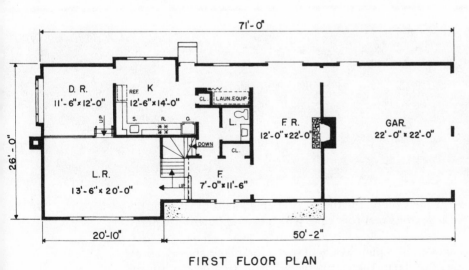

FIRST FLOOR PLAN

SECOND FLOOR PLAN

169

· SECOND FLOOR PLAN ·

The Knoll-Wood

Colonial elegance at its finest is provided by this two story design.

Gabled dormers, picturesque shutters, and a box bay window arrangement are but a few of the features that this traditional home offers.

Only a careful examination of the floor plans will reveal the numerous advantages that this home affords.

AREA: First floor   1,135 sq. ft.
        Second floor   990 sq. ft.

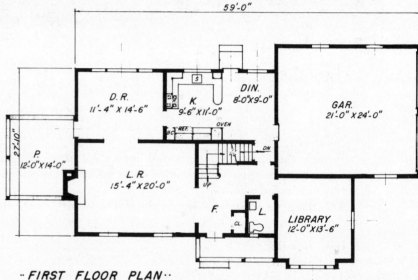

· FIRST FLOOR PLAN ·

The New Milford

One of the most interesting variations of the Colonial is the New England saltbox. It is distinguished from its sister-houses mainly because of its rear roof lines which extend downward much lower than the front roof lines.

Although the living room has four windows, it has plenty of wall space for imaginative decorating and features a brick-faced log burning fireplace. The combined kitchen-family room affords enough space for informal meals and relaxation.

Three of the four bedrooms are close to the family bathroom; the master bedroom has two sliding door closets, four windows and a glass enclosed shower stall and full-length mirrored dressing vanity.

This is a good, solid house without frills but with comfort and livability built in.

AREA: First floor   1,250 sq. ft.
        Second floor   1,078 sq. ft.

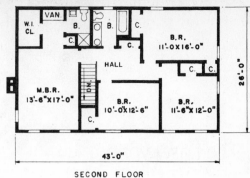

SECOND FLOOR

The New-Englander

The contemporary interpretation of this plan recalls the stately homes of New England with the emphasis on solid comfort and balanced proportions. Four bedrooms on the second floor complete a plan that includes a living room, dining room, kitchen-dinette and family room on the first floor.

AREA: First floor   1,162 sq. ft.
         Second floor   1,118 sq. ft.
         Total   2,280 sq. ft.

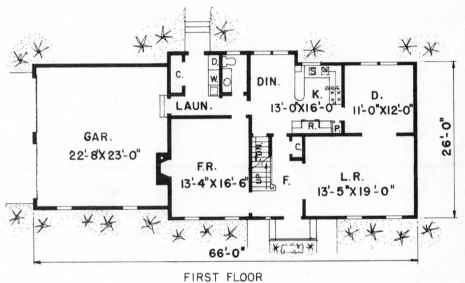

FIRST FLOOR

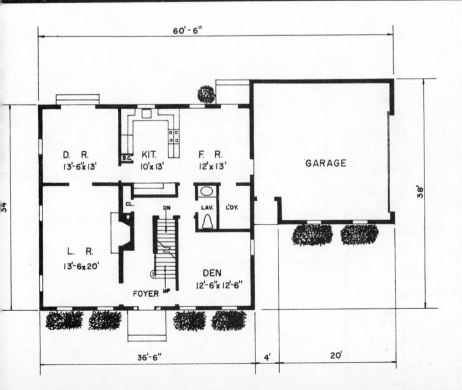

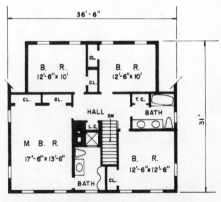

The East-Lynne

This home is complete and expected to satisfy all family demands right from the start while fitting a low building budget. Its four bedrooms make it a home that would be hard to grow out of. Room arrangement is the essence of efficiency and the rooms are of size and shape needed to handle their assignments. The constant demand of housewives for more closets is amply supplied.

Added formality is given to the living and dining rooms by the large and impressive main foyer, that is the key to circulation and makes it possible to distribute traffic effectively throughout the first floor areas and second floor bedrooms.

The combined family kitchen and dinette offer more than 15' of width across the back, enough to accommodate the entire family.

A few steps away are the laundry room and a convenient lavatory near the family room. The clean modern lines of the fireplace with its paneled walls dominate the rear wall of the family room.

All the bedrooms have cross ventilation and are liberally supplied with closets. The master bedroom has a private full bath with stall shower and an angular full wall mirrored vanity.

AREA: First floor   1,167 sq. ft.
         Second floor   959 sq. ft.

The Kirk-Wood

Inside this two story transitional home are many of the features most requested by those undertaking new home construction—wrapped in a pleasant, highly acceptable Colonial exterior of hand-split red cedar wood shingles and brick veneer. The sunken family room is separated from the kitchen-dinette by a wrought iron rail and features a beamed ceiling, stone fireplace and sliding glass doors.

A gentle stairway leads to the second floor master bedroom suite with a room size walk-in closet, two complete bathrooms and three additional bedrooms.

AREA: First floor   1,270 sq. ft.
         Second floor   1,204 sq. ft.
         Garage   517 sq. ft.

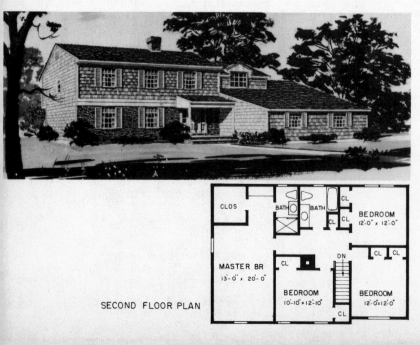

SECOND FLOOR PLAN

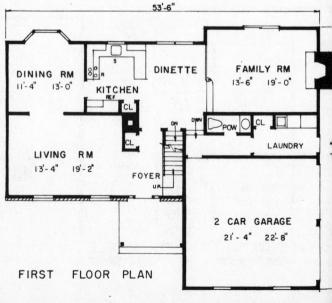

FIRST FLOOR PLAN

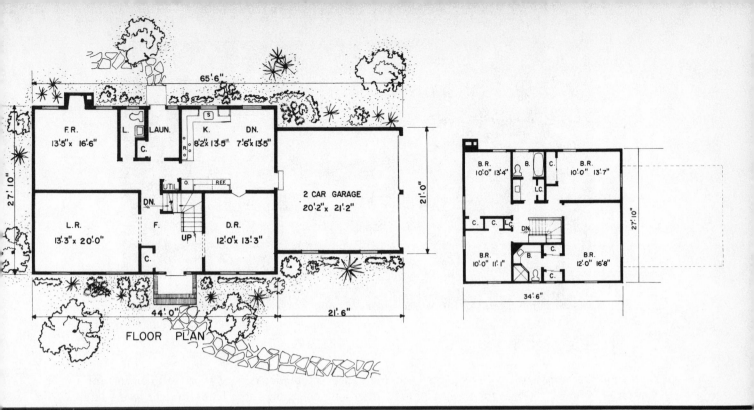

FLOOR PLAN

F.R. 13'5" x 16'6"
LAUN.
K. 8'2" x 13'5"
DN. 7'6" x 13'5"
2 CAR GARAGE 20'2" x 21'2"
L.R. 13'3" x 20'0"
UTIL
DN.
UP
D.R. 12'0" x 13'3"

65'-6"
27'-10"
21'-0"
44'-0"
21'-6"

B.R. 10'0" 13'4"
B.R. 10'0" 13'7"
B.R. 10'0" 11'1"
B.R. 12'0" 16'8"
DN.

34'-6"
27'-10"

The Devonshire

There is no doubt that the romance of English Tudor architecture is captured in the exterior styling of this two story four bedroom plan, which should delight families with a taste for continental design; the diamond-shaped leaded windows, half-timber and stucco evoke memories of the past, but the floor plan is strictly contemporary.

The front entrance is sheltered over the gabled overhang which leads into an impressive entry large enough to welcome guests and provides a handsome view of the fireplace and the beamed-ceiling family room on the left and the living room on the right.

AREA: First floor   1,333 sq. ft.
         Second floor   1,333 sq. ft.

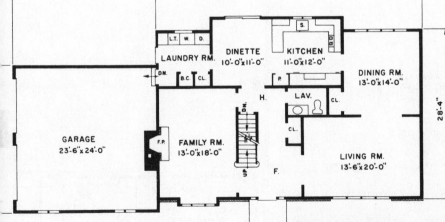

24'-6"
45'-6"

LAUNDRY RM.
DINETTE 10'-0"x11'-0"
KITCHEN 11'-0"x12'-0"
DINING RM. 13'-0"x14'-0"
GARAGE 23'-6"x24'-0"
FAMILY RM. 13'-0"x18'-0"
LAV.
LIVING RM. 13'-6"x20'-0"

28'-4"

FIRST FLOOR PLAN

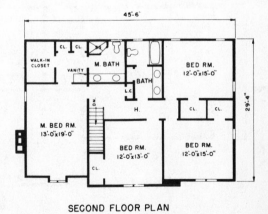

45'-6"

WALK-IN CLOSET
M. BATH
VANITY
BATH
BED RM. 12'-0"x15'-0"
M. BED RM. 13'-0"x19'-0"
BED RM. 12'-0"x13'-0"
BED RM. 12'-0"x15'-0"

29'-4"

SECOND FLOOR PLAN

The George-Towne

This charming home with its gabled entrance portico provides beauty as well as comfort for a large family . . . formal foyer separating living room and dining room leads straight to the large kitchen and family room with its angled fireplace . . . two baths service the three upstairs bedrooms . . .

AREA: First floor 1,317 sq. ft.
Second floor 1,317 sq. ft.

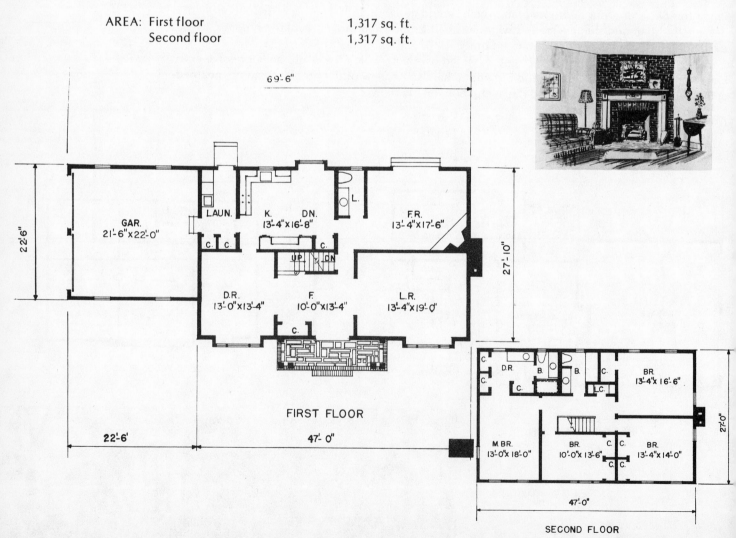

FIRST FLOOR

SECOND FLOOR

The Fair-Lawn

The elegance of a two story home will always remain despite the changing trends in home building.

This balconied eight room home displays that everlasting Southern charm of wrought iron and brick with a stately gabled wing on one side and a long roofed garage wing on the other. The finishing touches of cupola, fence, lamp post, and circular sash complete this picture.

Inside the planning has evolved a most modern arrangement of rooms for convenient and gracious living. A multitude of spacious closets, two and one-half baths, a delightfully large kitchen, dining room, living room, and den plus four amply sized bedrooms all combine to suit your mode of living.

AREA: First floor   1,270 sq. ft.
               (excl. garage & porch)
      Second floor   1,280 sq. ft.

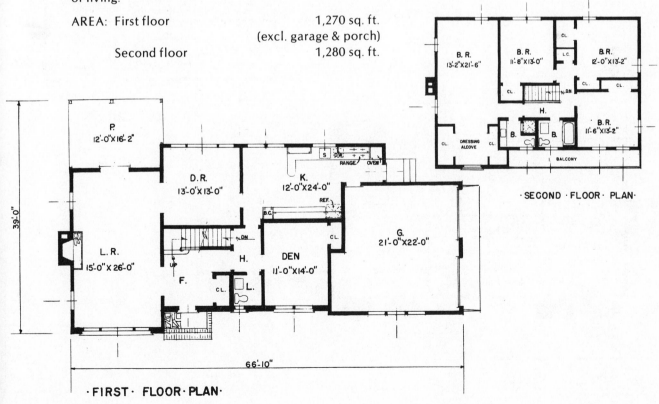

· SECOND · FLOOR · PLAN ·

· FIRST · FLOOR · PLAN ·

The Marquette

The capacity of a 17th Century French mansard roof to provide extra space on the second floor is clearly illustrated in this pleasant traditional house. Because the lower slope of the roof is very slight . . . the floor area upstairs is almost the same as the first floor. Old world charm emerges from such exterior details as the mansard roof, brick quoins at ends of structure, wood shingles on garage, scalloped leaded glass entrance doors and windows and paneled mouldings.

AREA: First floor   1,364 sq. ft.
       Second floor   1,264 sq. ft.
       Total   2,628 sq. ft.

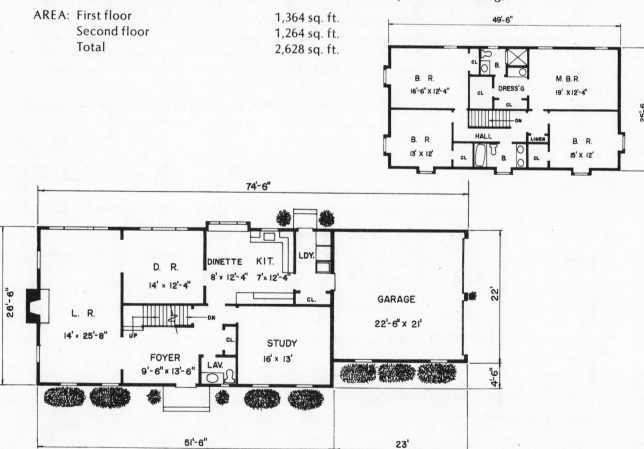

The Wickham-Woods

Few styles of residential architecture have the comfortable warmth of the houses built in America during the early 19th century. The secret lay in their simplicity. This design has many of the early earmarks—the big chimney, hand split wood shingles, small paned windows with shutters and paneled entrance door.

Inside, the plan shifts to the twentieth century with stepsaving physical comfort and convenient traffic control. The big kitchen-dinette has a full complement of cabinets, appliances plus a picture window dinette. A spacious foyer, living room, dining room, powder room, family room and laundry complete the first floor.

Upstairs are four bedrooms with ample closets, and two complete baths.

AREA: First floor   1,340 sq. ft.
       Second floor   1,070 sq. ft.
       Garage   590 sq. ft.

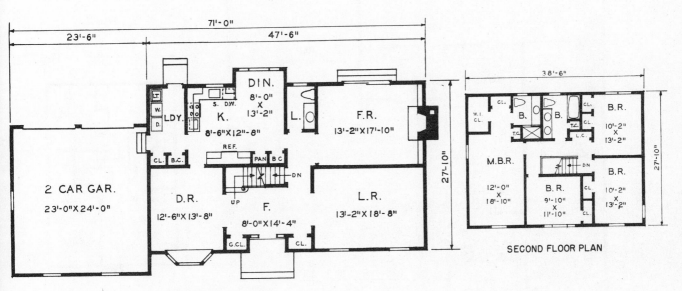

FIRST FLOOR PLAN

SECOND FLOOR PLAN

The Donny-Brooke

This New England "L" shaped colonial design with its exterior of wood shingles and vertical siding is surely an "eye-catcher."

From the entrance foyer there is ideal circulation to all principal first floor areas along with an open staircase leading to the four second floor bedrooms.

The master bedroom with private bath, dressing area and sit-down vanity, along with its large walk-in closet, becomes a suite of its own.

The women of the house will be overjoyed with the large family size kitchen that this plan offers. The dinette in itself is larger than many dining rooms and will easily seat six.

Connecting the kitchen and garage, as well as leading directly to the front and rear yards, is a spacious laundry room.

Note the wood paneled study with its own fireplace and closet. It is an ideal spot for relaxing by one and all.

AREA: First floor   1,345 sq. ft.
        Second floor   1,064 sq. ft.

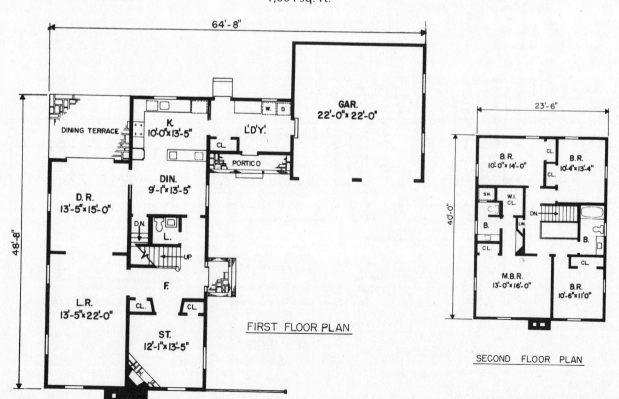

FIRST FLOOR PLAN

SECOND FLOOR PLAN

The Wood-Mont

Colonial charm could hardly be more appealingly captioned than this history-based design with a horizontal look. Traffic is effectively distributed throughout the first floor and by means of an attractive wrought iron staircase directly to the four second floor bedrooms.

AREA: First floor   1,340 sq. ft.
        Second floor   1,315 sq. ft.

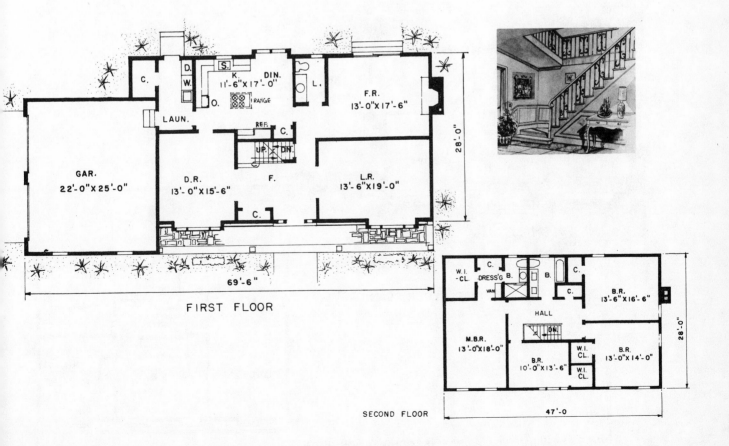

FIRST FLOOR

SECOND FLOOR

The Wood-Gate

This 18th century Dutch colonial exterior cloaks a 20th century floor plan.

Ideal traffic circulation is evident in this Colonial design. The large living room that measures 13' x 24' has three exterior exposures and colonial fireplace.

The laundry, located just a few steps from the kitchen and leading directly to the outdoors, with its closet and counter work area is a must in today's colonial homes. This area will surely satisfy the housewife's needs.

A second fireplace is located in the family room, which is accessible from the kitchen or dining room.

The large glass sliding doors tend to bring the outdoors in for that much desired "outdoor-indoor" living.

On the second floor, note the size (6' x 9') of the master bedroom walk-in closet with built-in shelves and shoe racks.

The remaining three bedrooms are all twin size with liberal closet space.

AREA: First floor   1,365 sq. ft.
        Second floor   1,030 sq. ft.

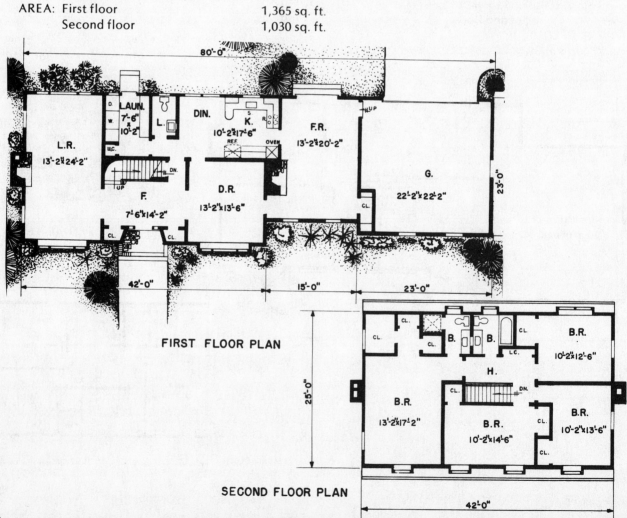

FIRST FLOOR PLAN

SECOND FLOOR PLAN

The Deer-Field

Ideal traffic circulation is evident in this Colonial design. The large living room that measures 13' x 24' has three exterior exposures and colonial fireplace . . . a "must" in today's colonial homes is the laundry, located just a few steps from the kitchen and leading directly to the outdoors, with its closet and counter work area . . . This area will surely satisfy the housewife's needs.

A second fireplace is located in the family room, which is accessible from the kitchen or dining room . . . The large sliding glass doors tend to bring the outdoors in for that much desired, "outdoor-indoor" living . . . On the second floor, note the size (6' x 9') of the master bedroom walk-in closet with built-in shelves and shoe racks . . . The remaining three bedrooms are all twin size, liberal closet space.

AREA: First floor   1,365 sq. ft.
         Second floor   1,050 sq. ft.

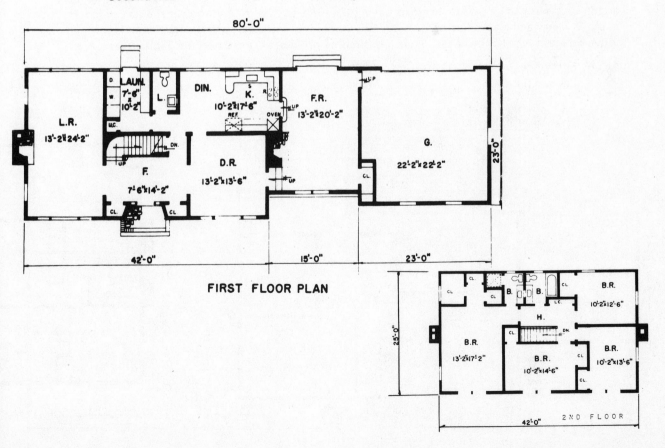

FIRST FLOOR PLAN

The York-Towne

Here is a home with exterior charm of the 18th century house of its type, but its floor plan fulfills the requirements of modern residential layouts. The much-desired good traffic pattern is evident, since one can go from the foyer to the kitchen, dining room, living room or family room on the first floor, or to the bedrooms on the second floor without crossroom circulation.

The combined kitchen-breakfast room arrangement offers more than 18 feet of width across the back of the house. Only a few steps away is the rear service entry with a mud-room closet adjacent to the laundry room.

The living room has wall areas well suited to various kinds of furniture arrangements as does the adjacent dining room. Off the entrance foyer are a lavatory at one end and, on the other, an open stair with a midway platform for well-lighted halls on both floors. Informality would likely be the order of the day in the family room, nearly 20 feet long, with pine-paneled walls, a brick fireplace and a large sliding glass door leading to the spacious outdoor terrace.

On the second floor, three bedrooms are close to the main family bath.

The fourth room, occupying the entire rear of the floor, is called a master bedroom suite by the architect. It has two large walk-in closets, a large private bath with tiled shower stall, a full-length counter vanity and a compartment enclosed water closet. Entrance into the bathroom is through louvered doors which may be left open, if desired, to provide a "dressing-room" appearance.

AREA: First floor   1,356 sq. ft.
           Second floor   1,060 sq. ft.

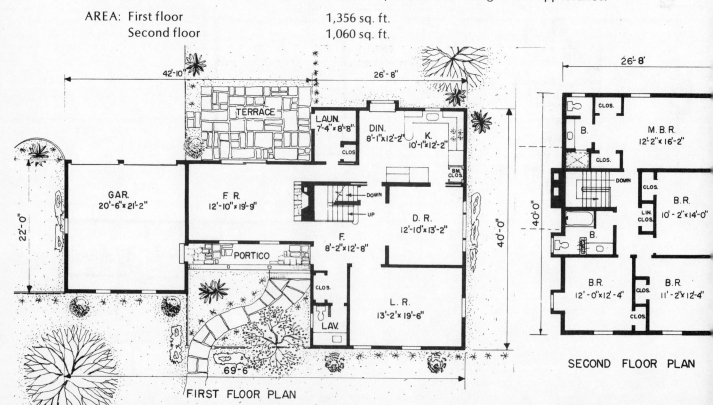

FIRST FLOOR PLAN

SECOND FLOOR PLAN

The Short-Hills

This is truly a southern colonial with its large portico and square columns.

The curved staircase leading to the second floor foyer, family room with incorporated "bar" and first floor den are only a few features that this plan offers.

A skillfully planned kitchen with window-walls surrounding the dinette affords a picturesque view while dining.

The master bedroom with rear window-wall and door leads to a sitting balcony overlooking the rear garden.

Completing the second floor, the remaining three large bedrooms and hall bath will satisfy any family needs.

AREA: First floor   1,371 sq. ft.
       Second floor   960 sq. ft.

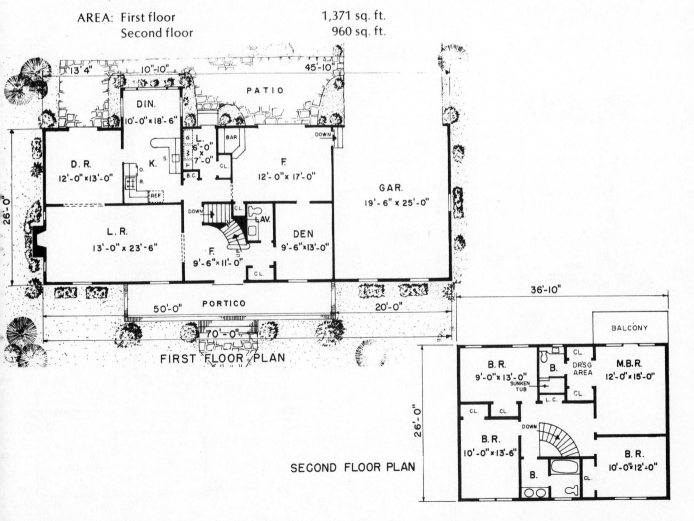

FIRST FLOOR PLAN

SECOND FLOOR PLAN

The Chevy-Chase

There are no fancy features to this two-story variation of an Early American house, but there is a definite feeling of solidity and warmth that suggests comfortable living and a rectangular design that cuts construction costs. Of special interest is the sizeable entrance foyer which is more than 14 ft. wide and makes a fine reception area. Traffic is effectively distributed through the first floor and by means of an attractive staircase directly to the four bedrooms on the 2nd floor. This house makes maximum use of every square foot of space on the inside and has old-fashioned charm on the outside.

AREA: First floor   1,375 sq. ft.
           Second floor   1,060 sq. ft.

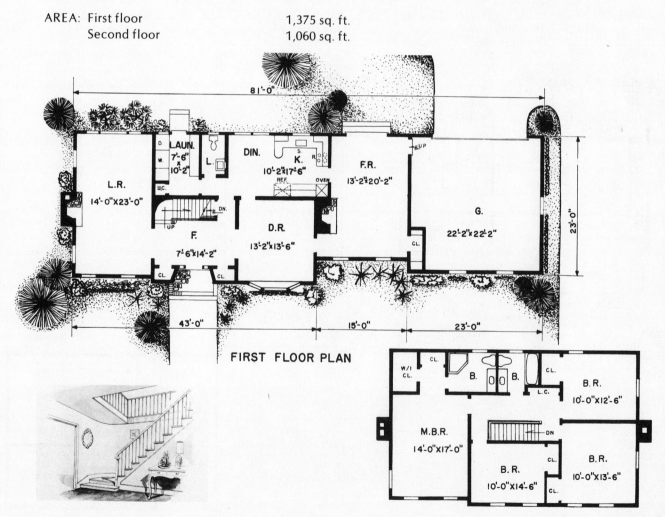

FIRST FLOOR PLAN

The Park-Lane

Here is an expandable two story colonial design.

The future fourth bedroom directly off the second floor foyer can also be used as a den, or children's playroom. This room can be finished with only minor expense as the need arises.

Within a few steps of all bedrooms is the main family "split" bath with toilet and tub separated from the vanitory area. This vanity area with wall to wall mirror and built-in planter and towel cabinet is unsurpassed.

Four living areas with interconnecting traffic flow, plus a conveniently located powder room and laundry, allow for excellent formal or informal entertaining.

A spacious basement with incorporated two car garage allows this home to be built on most of today's minimum sized building sites.

AREA: First floor   1,390 sq. ft.
           Second floor   918 sq. ft.

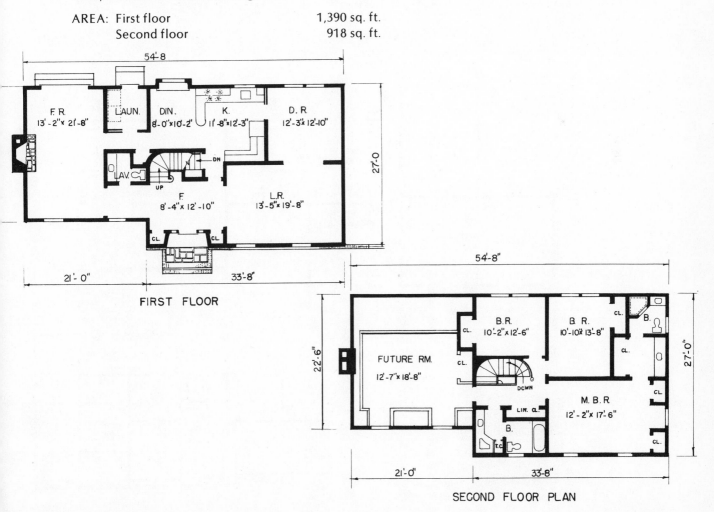

The Wimbledon

Once again, English architecture is enjoying wide popularity because there is something about the dark hand-hewn timber and stone exterior, the many paned and diamond shaped windows and the overall look of solidarity of this style that gives an impression of enduring comfort and happiness. Typical of English styling is the open staircase which leads directly from the entrance foyer to the four bedrooms and open balcony on the second floor.

A decorative metal circular staircase provides ready access to the upper "balcony library" that is located at the end of the living room, while directly behind is the "beamed ceiling" family room which connects with the outdoor terrace. Tasteful touches of Tudor styling suggest the relaxed living of this two-story design.

AREA: First floor   1,450 sq. ft.
       Second floor   1,500 sq. ft.
       Basement   1,200 sq. ft.
       Garage   560 sq. ft.

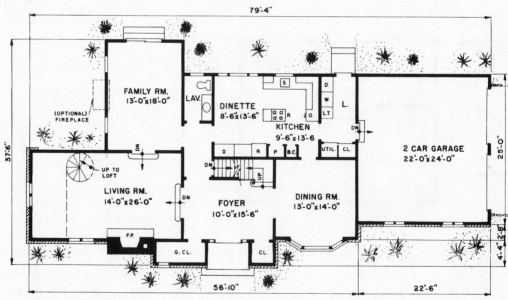

The Westminster

The Tudor adaptation of this three bedroom two-story design will make it stand out as a home of distinction. Its elegant exterior appearance is matched only by the quality of the interior design which was created for the modern family.

Inside, the unusually large foyer makes a fine reception area with its two coat closets and is the key to efficient circulation, distributing traffic effectively throughout the first floor and by an attractive staircase to the two bedrooms on the second floor. To the left is the formal dining room and the oak-paneled family room directly behind it features a stone fireplace with a raised flagstone hearth flanked with casement windows on either side.

There is no doubt that the romance and rustic charm of the English half-timber style of this three bedroom two-story design should delight families with a taste for continental design.

AREA: First floor   1,458 sq. ft.
         Second floor   539 sq. ft.
         Basement   1,458 sq. ft.
         Garage & Laundry   639 sq. ft.

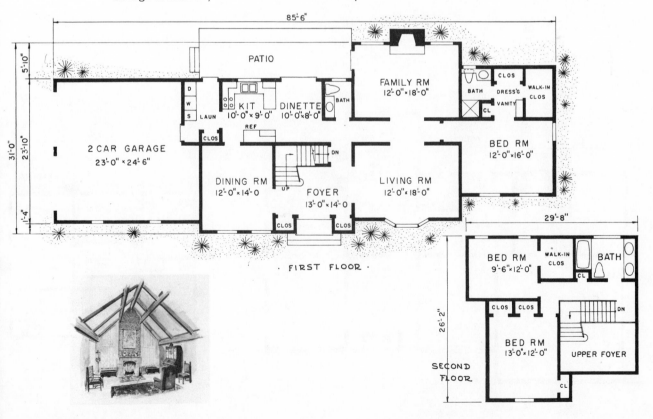

· FIRST FLOOR ·

SECOND FLOOR

The Old-Field

Styles in houses may come and go, but for enduring popularity it is hard to beat the familiar two-story Colonial.

Inside, the large and impressive foyer makes a fine reception area as well as being the key to efficient circulation, distributing traffic effectively throughout the first floor and by means of a grand circular staircase directly to the second floor bedrooms.

For the all-around privacy and economy of two stories, plus all the conveniences of colonial living, this home is ideal for a large or growing family and exudes the comfort and warmth inherent to traditional designs.

AREA: First floor   1,510 sq. ft.
       Second floor   1,190 sq. ft.
       Total   2,700 sq. ft.

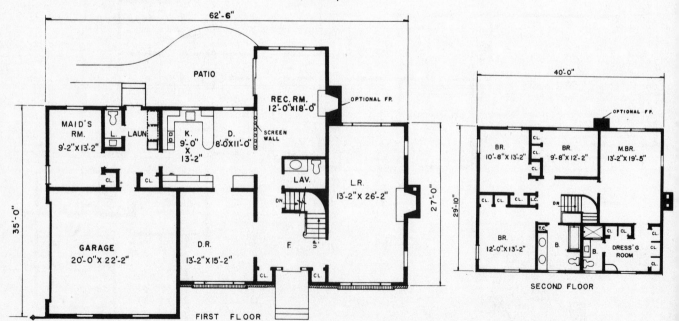

The Marc Woods

Because of the great circulation pattern, you can go to the foyer, kitchen, dining room and living room without crossing through other rooms on the way . . . both the living room and family room have a fireplace . . . four bedrooms and optional fifth bedroom or den are on second floor as well as two baths . . . living area is:

AREA: First floor   1,514 sq. ft.
          Second floor   1,188 sq. ft.

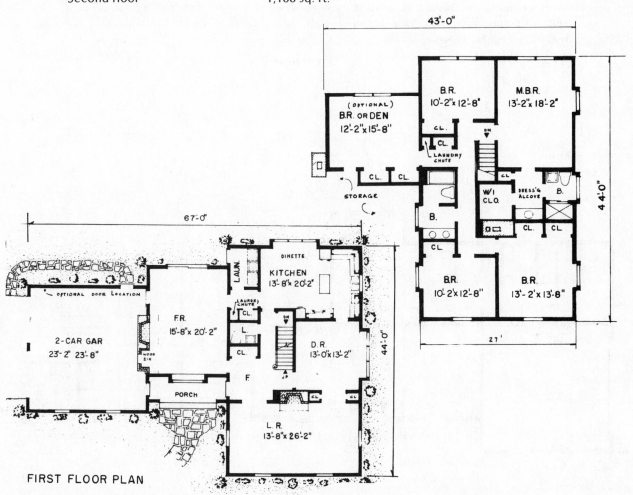

FIRST FLOOR PLAN

The Bain-Bridge

The twenty-six foot long living room with its brick fireplace and large picture window at both ends offers unlimited furniture arrangements. Note the spacious recreation room adjoining the kitchen-dinette with its "U" shape work center, conveniently located between the recreation room, dining room and dinette. The maid's room can also double as a guest room. The lavatory with adjoining laundry provides an excellent "inside-outside" utility area, also containing a pantry closet and connecting directly to the 2 car garage. The second floor provides four king size rooms including two baths and dressing room with five closets for the master suite.

For all around privacy of two stories plus all the conveniences of colonial living, this home is ideal for a large, or growing family.

AREA: First floor   1,510 sq. ft.
       Second floor   1,190 sq. ft.

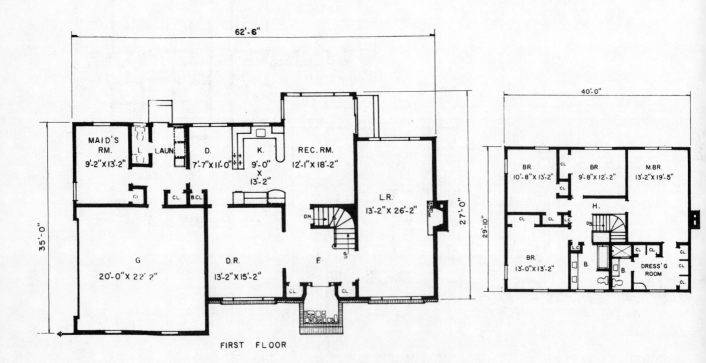

FIRST FLOOR

The Stone-Haven

This interesting Dutch Colonial features the space-creating gambrel roof. The sunken living room has three exposures with wrought iron railings producing a balcony effect to the dining room. An optional additional room on the second floor can be used as a den, sewing room or 5th bedroom.

AREA: First floor   1,514 sq. ft.
        Second floor   1,188 sq. ft.

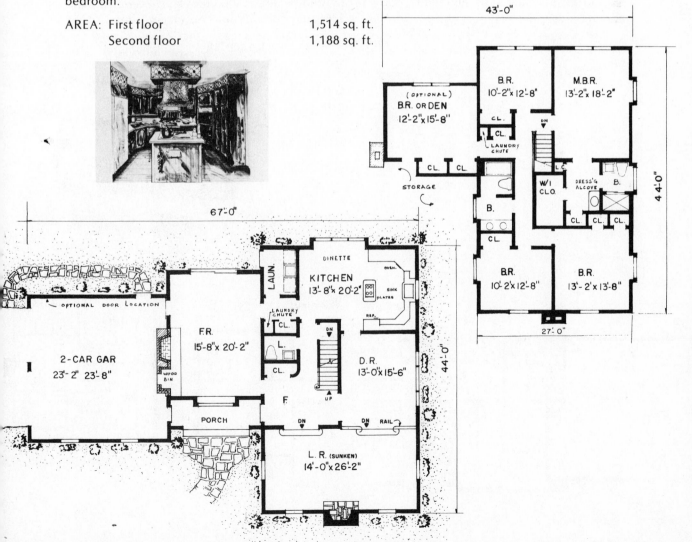

The Wake-Field

Authentic detailing of brick arches, leaded glass doors and windows, brick quoins, hip roof and massive chimney reflect the taste and traditional character of the ever popular French Provincial style. The exterior of this two story, three bedroom design is reminiscent of the country estates of the by-gone era, while the interior combines the formal with the casual for a truly comfortable home for the entire family.

Inside the recessed double doors, the large foyer makes a fine reception area with an impressive staircase and wrought-iron handrail leading to the two bedrooms on the second floor.

Although the exterior of this inviting house recalls another era and should delight families with a taste for continental design, the interior is for modern day comfort and convenient living.

AREA: First floor   1,514 sq. ft.
       Second floor   536 sq. ft.
       Basement   1,580 sq. ft.
       Garage & Laundry   618 sq. ft.
       Patio   250 sq. ft.

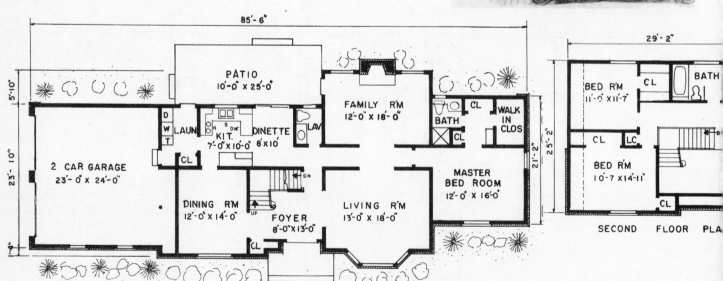

FIRST FLOOR PLAN

SECOND FLOOR PLAN

The East-Windsor

The colonial dignity of the exterior of this stately two story home carries throughout the planning of the interior. Spaciousness is a keynote here, and is apparent from the room sizes indicated.

The service area here is a housewife's dream. Separate dining and food preparation areas in the kitchen and a large utility room with counters, washing equipment, freezer and ironing. A service lavatory right next to the rear entrance and a separate stair to the maid's room and bath located over the garage.

Four large bedrooms with walk-in and sliding door closets plus two full baths make up the 2nd floor sleeping area of this home.

AREA: First floor   1,627 sq. ft.
                (excl. garage & porch)

Second floor   1,510 sq. ft.

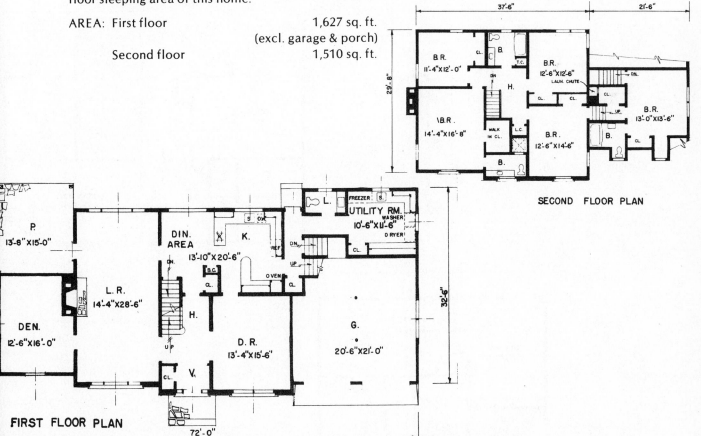

SECOND FLOOR PLAN

FIRST FLOOR PLAN

The Somerset

The traditional appearance of this southern colonial with its large portico and square columns reflects the comfort and convenience built within and makes it right at home in the city, suburb or country.

The air of gracefulness is immediately apparent as one moves under the entrance portico, which shelters the front entrance door. What better welcoming sight than a large foyer with a sweeping grand circular staircase? This is the key to the efficient circulation that permits effective traffic distribution throughout the first floor areas and four bedrooms located on the second floor. The open spacious stairwell with its wrought iron ornamental handrailing eliminates any feeling of congestion on the second floor.

This is a house that makes maximum use of every square foot of space and has traditional old-fashioned enchantment, charm and dignity.

AREA: First floor   1,525 sq. ft.
        Second floor   1,185 sq. ft.
        Basement   1,525 sq. ft.
        Garage   630 sq. ft.

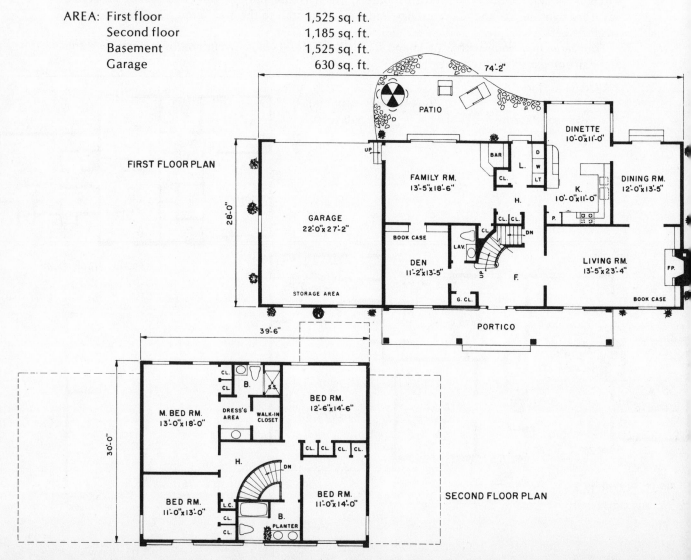

The Drift-Wood

All the regal luxury of the traditional two-story home is contained in the colonial styling of this large nine room plan. The front of this home presents long low-appearing lines. Faced with brick veneer, it is lengthened by the planting beds on either side of the entrance. Satisfying weather protection is afforded at the entrance and convenient access is afforded to the basement. Closet space is abundant and baths and lavatory are very well located and ample sized. For the all around privacy of two stories plus all the conveniences of colonial living, this five bedroom home is ideal for the large or growing family.

AREA: First floor   1,325 sq. ft.
         Second floor   1,600 sq. ft.

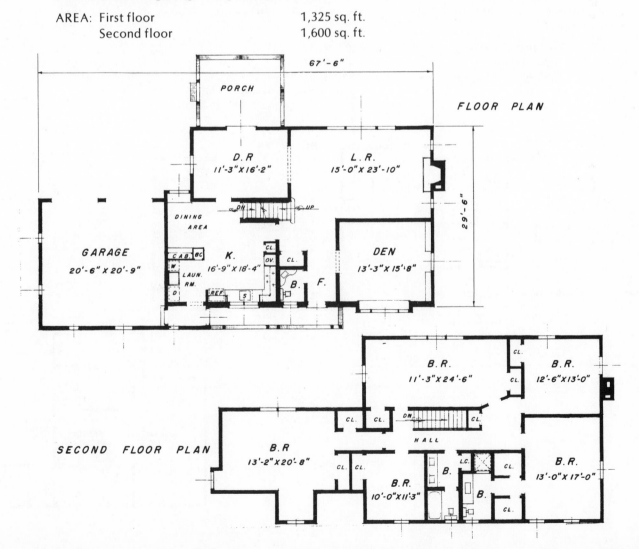

The Heritage

The formal appearance of this two story symmetrical Colonial design is softened by the addition of the barn-type two-car garage at one side and more than equalled in the spacious well-ordered interior room arrangement. The grand entrance foyer provides excellent circulation throughout the entire house—and of special interest are the twin fireplaces, one each in living and family rooms. For those with a yen for luxury, the second floor master bedroom suite offers two walk-in closets, a tub and stall shower bath, vanity in dressing area and a private sun-deck. A compartmentalized bath, with twin basins and full-wall mirrored vanity, services the other three bedrooms.

AREA: First floor   1,680 sq. ft.
        Second floor   1,473 sq. ft.
        Garage   575 sq. ft.

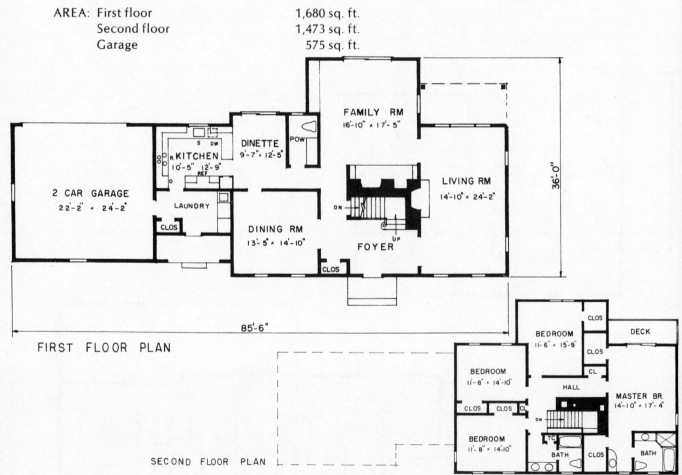

FIRST FLOOR PLAN

SECOND FLOOR PLAN

The Cedar-Brook

For the comfort-loving modern family here is a handsome choice that cherishes the heritage of its earlier ancestors. The Colonial paneled door opens to the large foyer, with sliding door coat closet and attractive stairway with wrought iron railing. Immediately to the right is the living room with an informal family room at the rear, with fireplace and sliding glass doors to the terrace.

The kitchen-dinette is a true family center. Off the kitchen is the laundry with its service entrance and a two car garage.

On the second floor are four lovely bedrooms with ample closets and two modern luxurious baths. One is in the master bedroom off a dressing area with a walk-in closet and built-in vanity. A centrally located powder room off the foyer serves the first floor, and the stairway to the full basement is in this central location for convenience.

AREA: First floor   1,720 sq. ft.
        Second floor   1,720 sq. ft.
        Garage   700 sq. ft.

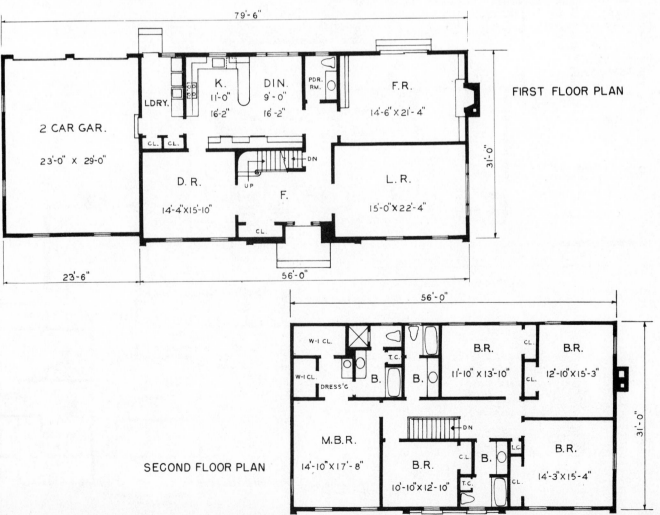

The East-Brook

There is no doubt that modern living inside and the romance of the traditional French Provincial countryside exterior styling of this two story four bedroom plan should delight families with a taste for continental design.

The circular staircase with wrought iron railing provides a luxurious access to the four bedrooms on the second floor that completes a plan which retains all the good living qualities and hospitality of an earlier era.

AREA: First floor   1,725 sq. ft.
        Second floor   1,562 sq. ft.
        Garage   576 sq. ft.
        Basement   1,725 sq. ft.

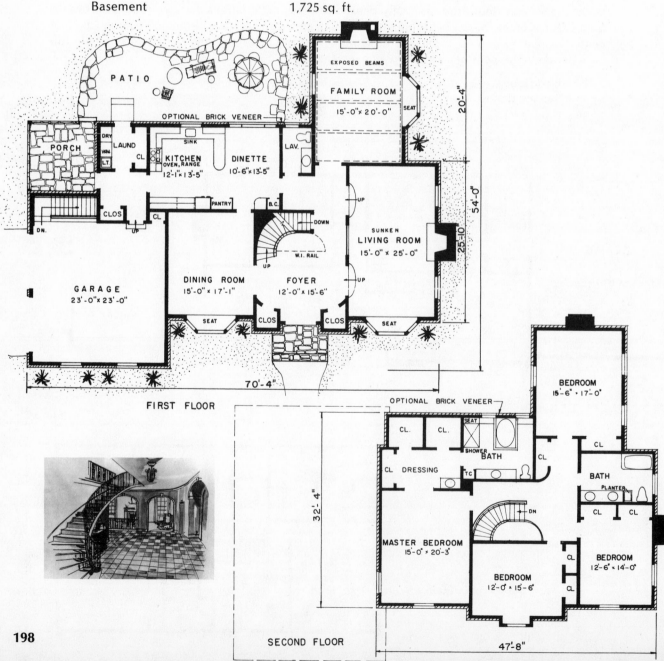

The Chateau-Blanc

Visions of royal living come quickly to mind in looking at this elegant French Provincial two story, four bedroom design. There's royal living inside, too, with spaciousness the keynote; the foyer is room size with a sweeping, curved staircase to the second floor; the kitchen features a dinette with a rectangular bay window, and the sunken family room at the rear is serviced by the raised entrance foyer level.

The multi-glass paned exterior French doors of the living and dining rooms open into the front balustrated court-yard.

A service stair to the basement from the two-car garage provides additional storage area, and extensive use of brick veneer helps to minimize maintenance requirements.

On the second floor the master bedroom suite features two closets, one a walk-in, and a complete bathroom with a full-length mirrored vanity; the two other bedrooms share a lavish bath and the fourth bedroom has a private bath.

AREA: First floor   1,840 sq. ft.
       Second floor   1,640 sq. ft.
       Garage   650 sq. ft.

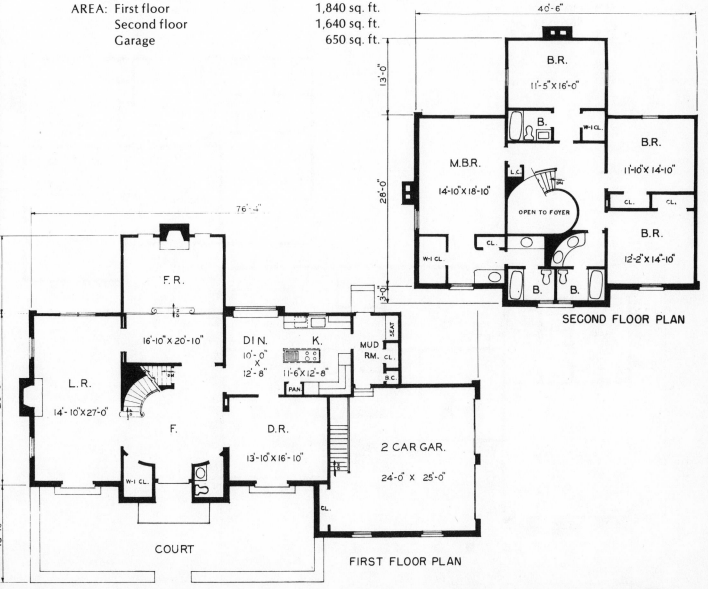

SECOND FLOOR PLAN

FIRST FLOOR PLAN

199

The Fair-Oaks

Here is a design with a French connection.

Although the day of French nobility and the historic period are gone, French Provincial styling is gaining renewed popularity in America; this two-story four bedroom model is a good example.

It is complete, with steep hip roof, charming window detailing that includes half-dormered windows, brick veneer with brick quoins at the corners, massive chimney, double front entrance doors and a decorative cupola on the roof of the garage wing.

There's something about the Continental elegance which makes homes fashioned in the romance of French architecture a little more livable, a bit more comfortable and a lot more lovable.

AREA: First floor   1,854 sq. ft.
          Second floor   1,596 sq. ft.
          Garage   485 sq. ft.

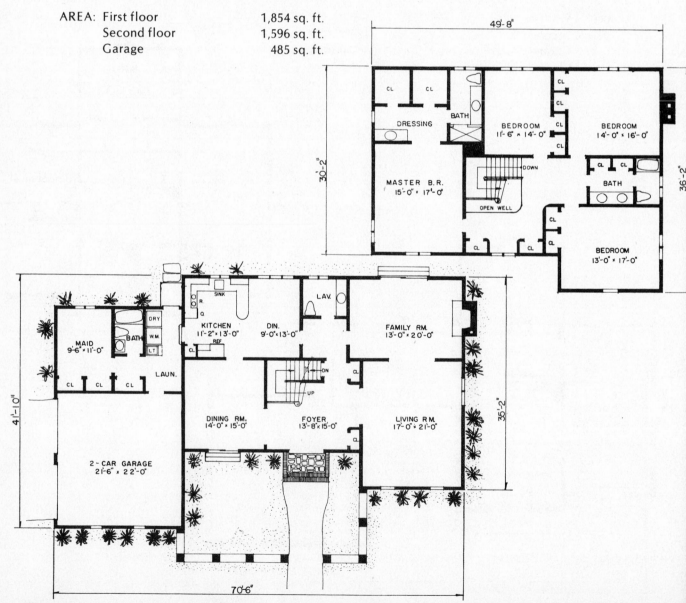

The San Mateo

Many details typical of the Southwest Spanish architecture are included in this two story four bedroom house. Stucco, brick veneer, arched and vertical windows, cantilevered balconies and hand carved wood double entrance doors. To the right of the entrance "room size" foyer is the living room, and behind it is the sunken family room with wood-beamed ceiling and a rubblestone wood-burning fireplace. Laundry facilities are located next to the kitchen-dinette and adjacent to the maid's room and the double garage.

A powder room is conveniently located just off the foyer.

The second floor is reached by the impressive curved staircase. The master suite has two closets, one a "room-size" walk-in, sitting area, open balcony, and a luxurious bath with twin-basin vanity, stall shower and Roman whirlpool bathtub. Two complete bathrooms service the other three bedrooms.

AREA: First floor   1,940 sq. ft.
        Second floor   1,620 sq. ft.
        Garage   620 sq. ft.

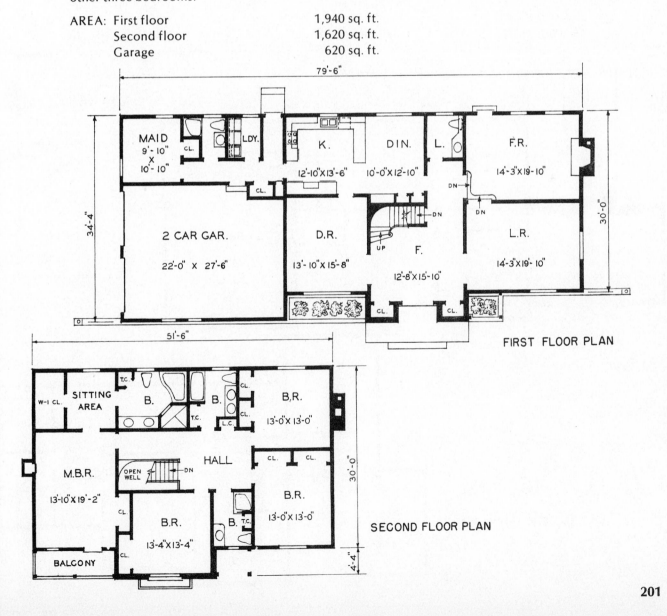

FIRST FLOOR PLAN

SECOND FLOOR PLAN

The South-Hampton

A palatial foyer leading to the sunken living room sets the tone for luxurious living in this elegant two story English Tudor design. Projecting bay and dormer windows, half-timber decorative work, smooth stucco, diamond paned windows and large chimney are authentic period details.

The angular dinette that opens to the rear deck is the dramatic feature of the kitchen-dinette area. The "cathedral ceiling" family room with brick fireplace also leads to the open deck. A laundry room, pantry, lavatory, powder room, and a two car garage with a "room-size" storage area complete the first floor.

Upstairs, the second floor contains a total of five bedrooms. The master suite is serviced by a private bath consisting of a twin-basin vanity, stall shower, two walk-in closets and dressing area. The other four bedrooms are convenient to the compartmentalized bath. An open balcony overlooks the family room below.

AREA: First floor   2,120 sq. ft.
        Second floor   1,950 sq. ft.
        Garage   660 sq. ft.

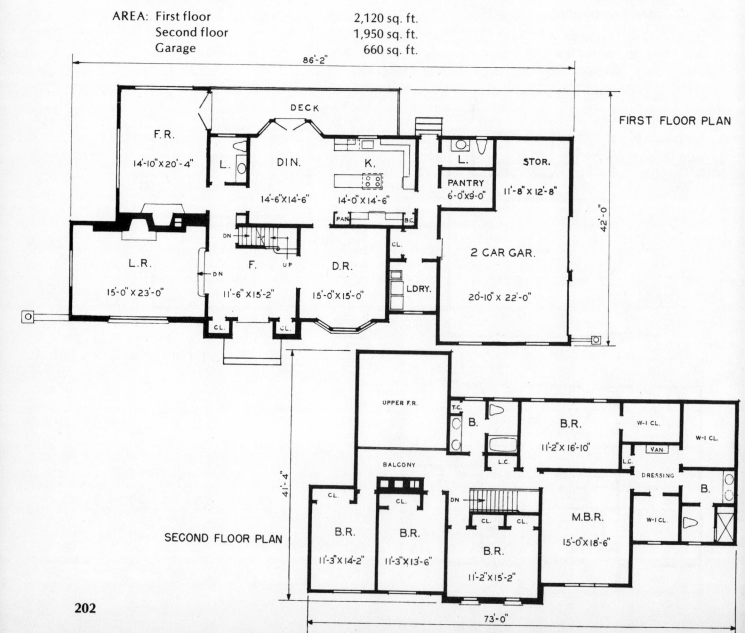

The Wellington

This impressive exterior with its stone veneer, half-timber, stucco, bays, half-dormers and diamond-paned windows is distinctively English Tudor. Visual variety, so pleasing outside, is continued indoors as well. The grand foyer forms the primary entrance and directs traffic into all parts of the house. Straight ahead the 17' × 23' living room with its cathedral ceiling features a massive brick "see through" fireplace and leads into the dining and family rooms. The kitchen-dinette is amply supplied with base and wall cabinets, appliances, etc. and equipped with an "island" range. The lavish private master bedroom suite on the first floor is accessible through a lounge, has three closets, one a walk-in, dressing room vanity and complete bath with glass enclosed stall shower. Three large bedrooms, two baths and a lounge or hobby room complete the second floor.

AREA: First floor   2,310 sq. ft.
         Second floor   1,146 sq. ft.
         Sundeck   141 sq. ft.
         Garage   644 sq. ft.

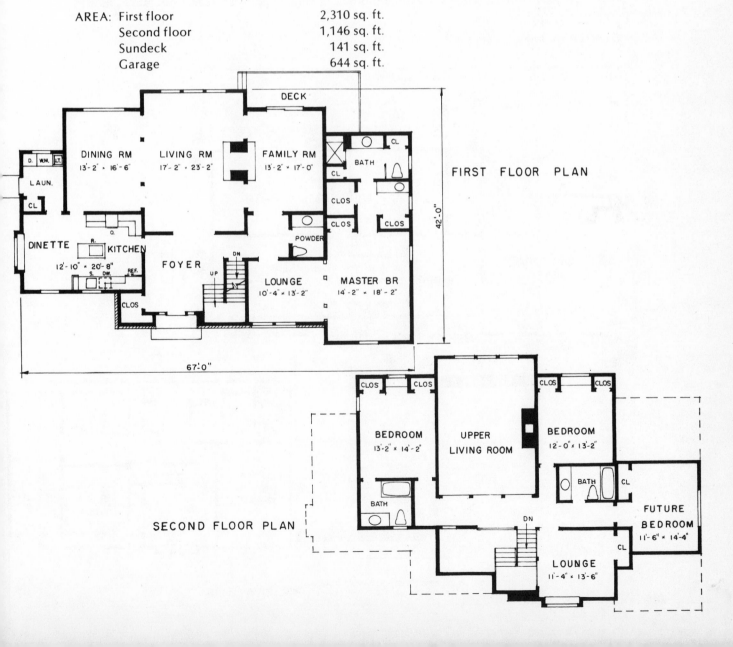

The New-Castle

Designed to provide esthetic appeal as well as complete comfort for large or growing family . . . low building cost . . . four spacious bedrooms . . . formal foyer separating living room from dining room . . . angled corner fireplace in family room . . . sliding glass doors lead from family roof to patio.

TOTAL LIVING AREA 2,440 sq. ft.

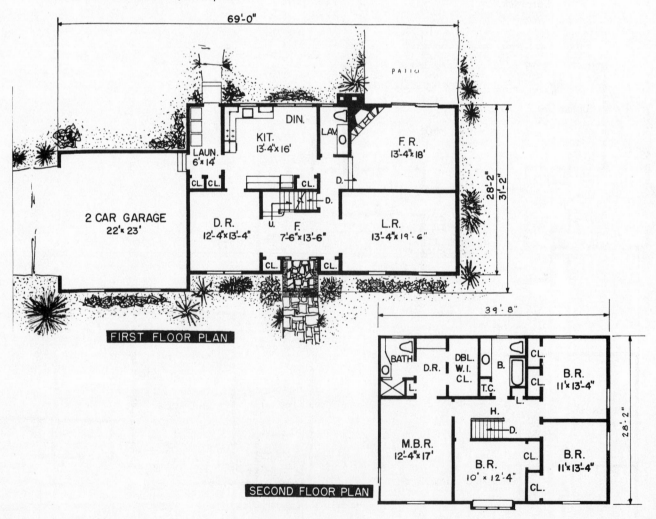

FIRST FLOOR PLAN

SECOND FLOOR PLAN

The Cortland

The drama of high sloped ceilings makes this five bedroom contemporary home something special. The living room and family room that feature a "see through" fireplace have beamed ceilings that soar to the height of the roofline, echoing the intriguing "slant" of the predominant exterior. Separated for complete privacy and quiet from the living area are the two bedrooms and bath off the second floor. Three additional bedrooms and bath off the second floor balcony that overlooks the family and living rooms below may be finished at a later date, if so desired.

An interesting feature is the rear L-shaped wood deck that is serviced by the sliding glass doors of the dinette, living and family rooms.

AREA: First floor   2,302 sq. ft.
        Second floor   888 sq. ft.
        Sun Deck   472 sq. ft.
        Garage   719 sq. ft.

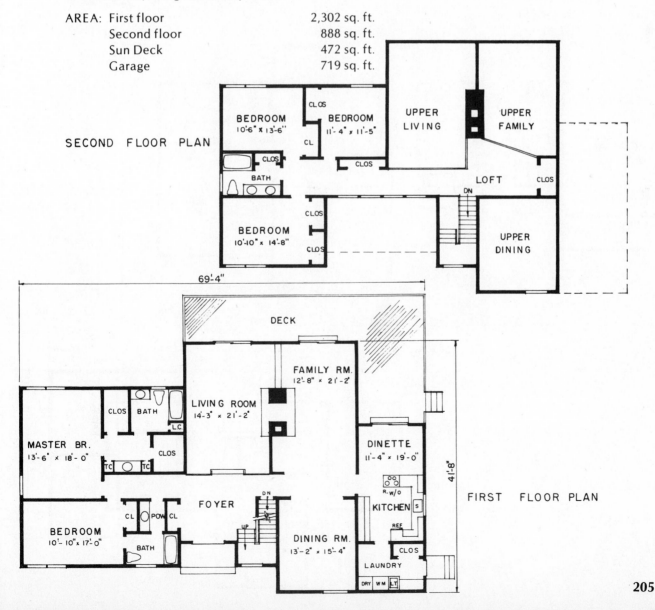

205

The Hawthorne

The impressive facade of this traditional design is achieved with the Colonial details of its one-story columned portico, hand split wood red-cedar shingles, small paned windows and the formal balance of the three second floor dormers and the first floor bay-windows. Inside—the grand foyer is the key to efficient circulation. Special features include a sunken living room; family room with fireplace, wet bar, and sliding glass doors to rear patio; island kitchen; two car garage with closed-off storage area; a private bedroom suite with five closets, complete bath with tub, stall shower, and twin basin full-wall mirrored vanity. Three bedrooms and two baths complete the second floor. The enduring look of this traditional design will be a source of pride in any neighborhood.

AREA: First floor   2,396 sq. ft.
         Second floor   637 sq. ft.
         Garage   616 sq. ft.

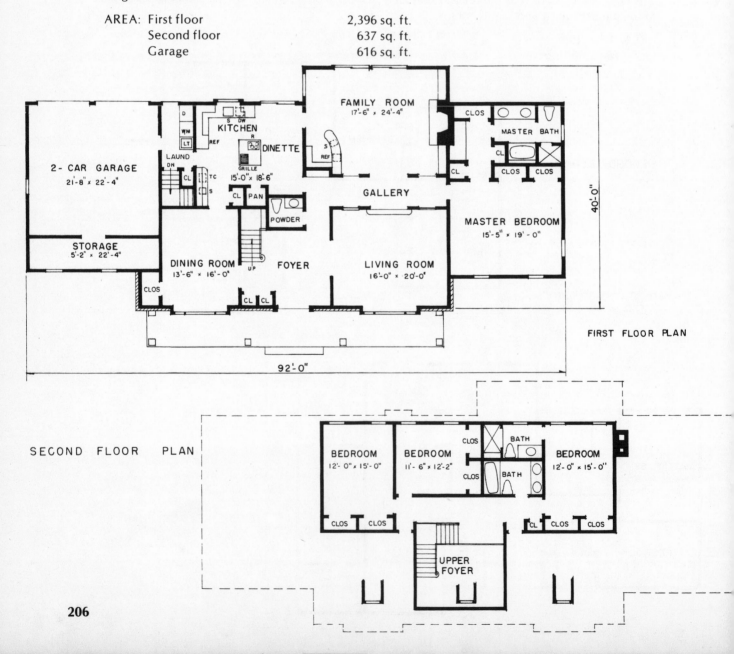

The Berkshire

The glamour and serenity of French Provincial styling are on display throughout this unusual 1½ story design and make this the perfect home for the family with a well-developed feeling for traditional influence. The main interior features are the grand foyer with its circular staircase that flows up to the two second floor bedrooms; the sunken living room that is flanked with wrought-iron rails and grilles; and the built-in wet bar and fireplace with French mantel in the living room. The master bedroom suite has two closets, one a room-size walk-in, dressing area vanity with full-length mirror and a complete bath with double-basin vanity, sunken Roman tub, glass enclosed shower and sauna. Accessible to the rear sun-deck are the living room and well-equipped "island" kitchen.

AREA: First floor   2,497 sq. ft.
        Second floor   527 sq. ft.
        Garage   550 sq. ft.

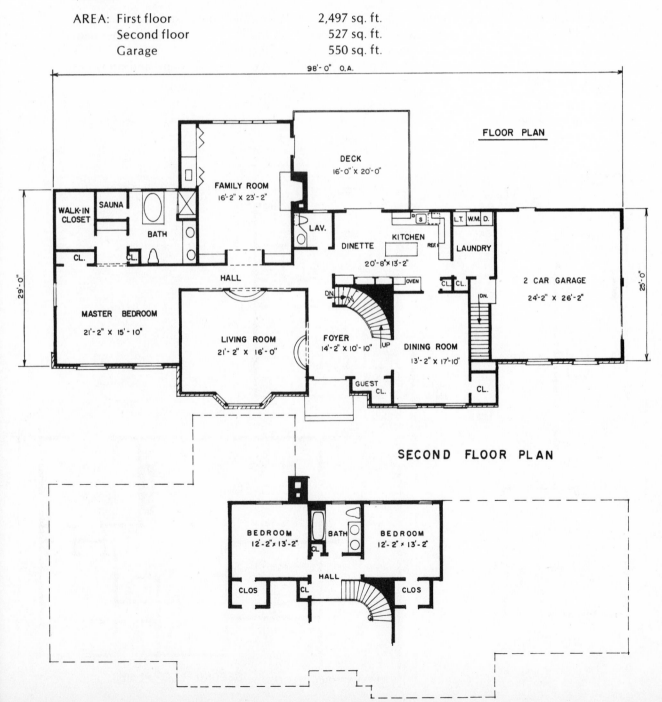

The Buena-Vista

Twentieth-century living at its most luxurious is possible in this unusual contemporary two-story design that features a "great room." The exterior is vertical V-joint redwood, asphalt shingle roof, casement and transomed windows. Weather-protected entry with glass side panels admits abundant light to the spacious foyer. Inside, to the rear is the dropped "great room" with its fireplace, full glass walls and sliding doors that open to the L-shaped wood deck; the U-shaped "cathedral ceiling" kitchen puts everything within easy reach, and, for informal family meals, there is the dinette, which has access to the rear deck.

To the right of the foyer is the bedroom area. At the rear is featured the master bedroom suite with its own private balcony, two walk-in closets, stall-shower and tub bathroom. The other bedroom has a double purpose compartmentalized "powder-bathroom."

Upstairs, the open foyer looks down on the "great room," and the two bedrooms and study are serviced by a common bath.

AREA: First floor   2,560 sq. ft.
         Second floor   1,130 sq. ft.
         Garage   570 sq. ft.

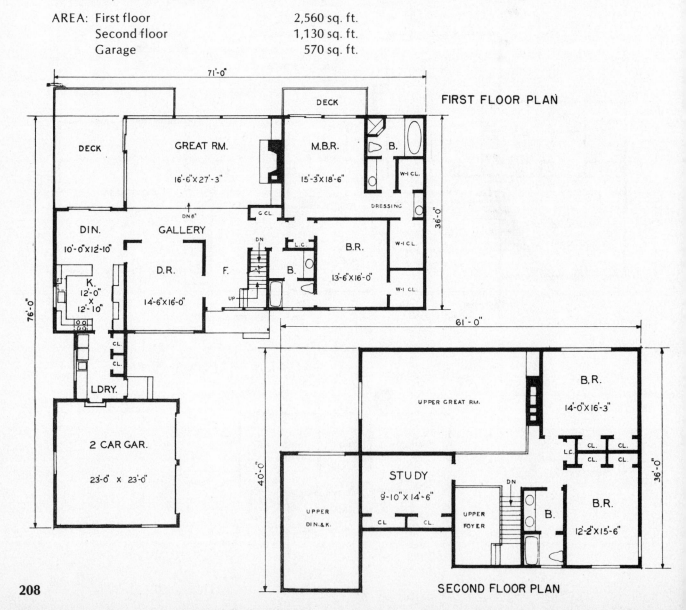

FIRST FLOOR PLAN

SECOND FLOOR PLAN

The Nottingham

This four bedroom, two story, full basement design has all the traditional elements of the English Tudor manor house. Basic characteristics feature a massive brick chimney with protruding chimney pots, steep roofs of varying heights, an angular bay window, narrow diamond-paned windows with leaded glass, and half-timber on stucco walls.

A laundry-mud room next to the kitchen has access to the two-car garage and the rear. Besides two closets in the laundry room there is enough space in the garage to take care of such things as garden tools, toys, lawn chairs and the like.

An open staircase leads directly from the entrance foyer to the sleeping area and the wide upstairs hall features a balcony with ornamental wood rail overlooking the living room below.

The tasteful touches of traditional styling and the imposing exterior of this Tudor home suggest the comfortable living that it offers.

TOTAL LIVING AREA: 2,834 sq. ft.

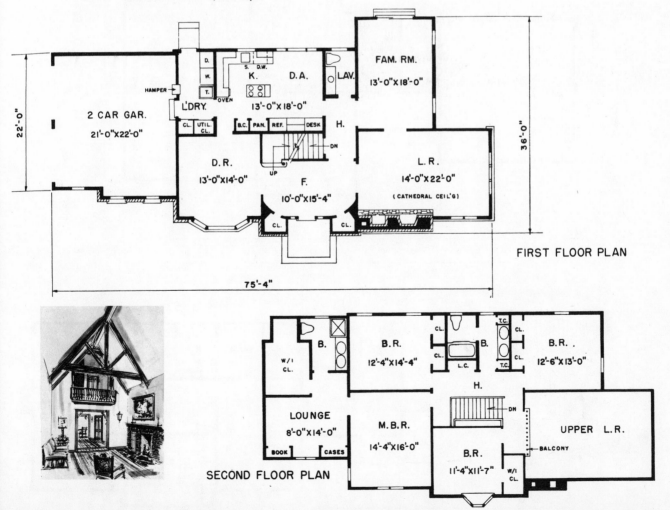

FIRST FLOOR PLAN

SECOND FLOOR PLAN

The Chateau-Gaye

Eye-catching exterior features the steep hip roof, curved window heads, wall-faced dormers, and diamond pane windows over the recessed entrance . . . living and family rooms have fireplaces . . . huge foyer has two closets . . . laundry room is off kitchen . . . four large bedrooms each with loads of closet space . . . luxurious lounge is off master bedroom.

TOTAL LIVING AREA: 3,175 sq. ft.

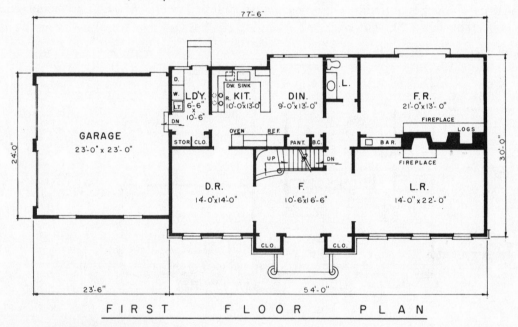

FIRST FLOOR PLAN

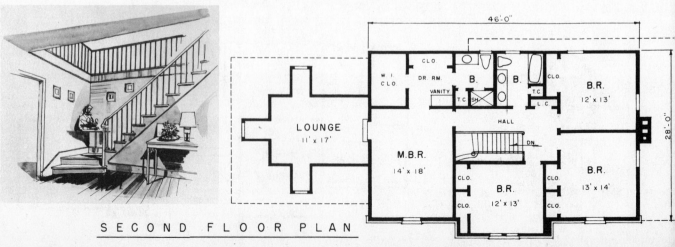

SECOND FLOOR PLAN

VACATION AND LEISURE-TIME HOMES

Whatever your taste, whatever your budget, the following designs for vacation or leisure-time living offer a change from everyday patterns. Today—more than ever before, Americans are investing in the future in a "second" home—it pays dividends in pleasure and relaxation, while increasing in value over the years.

Whatever your choice, the following designs will intrigue your imagination and complement your budget.

The Hilltop

The A-frame design, synonymous with vacation living, has received many architectural treatments. Here is another variation which proves that there is something new under the sun—25' x 25' square floor plan with a prow-shaped two-story glass expanse facing a wrap-around wood sun deck to take full advantage of your favorite view. The living area is quite dramatic with an exposed wood beam cathedral ceiling, a large rugged stone-wall corner fireplace, raised brick hearth with a dome-shaped hood, plank flooring, and sliding glass doors accessible to the deck.

Whatever your motive, whether it be a retreat in the woods, a cottage on the lake or a beach house by the shore, this design is all "decked out" for convenient living and complete relaxation.

AREA: First floor   625 sq. ft.
        Second floor   450 sq. ft.
        Deck   675 sq. ft.

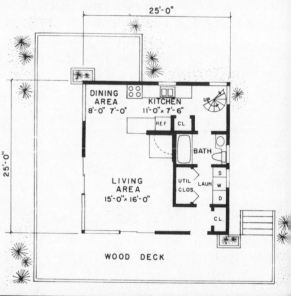

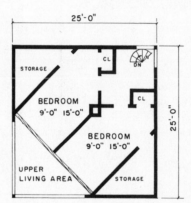

The Vacation Home

There is a great deal of eye appeal to the exterior of this very delightful A-frame design. The wrap-around, partially screened-in, sun drenched deck gives a choice of location for relaxing or entertaining from the large visual area of the second floor which serves as the living area. This is a basementless plan that has an electric heating unit for year-round living and has three bedrooms and two baths on the first floor.

AREA: First floor   676 sq. ft.
        Second floor   676 sq. ft.
        Sundeck: 356 sq. ft.
        Screened Porch   125 sq. ft.

The Camelot

Exciting exteriors, like this unique recreation house which suggests the Far East, sometimes can't be "themselves" in the suburbs, but come to life and are perfect for a picturesque building site in the woodlands or at the beach.

Whatever your motive, whether it be a retreat in the woods, a cottage on the lake, or a beach house by the shore, this design, all "decked out" with an oriental flavor, will answer your need for relaxation and that get-away-from-it-all leisure feeling

AREA: Living level   702 sq. ft.
          Bedroom level   702 sq. ft.

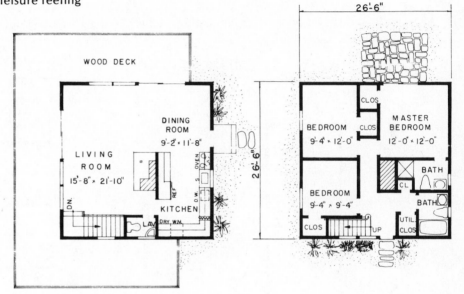

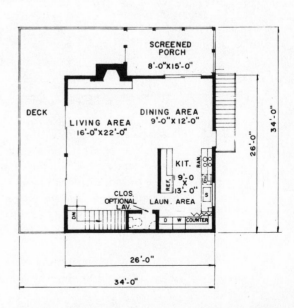

The Sea-Girt

This simple contemporary vacation house of diagonal V-joint red wood siding is ideally suited for a wooded setting. It is a vacation house that contains all the basic elements of leisure living, without many of the frills and luxuries you might want in your year-round home.

The first floor contains a living and dining area with massive windows overlooking the scenery in all directions.

If a vacation house is in your plans, consider the long range economy and comfort of this design.

AREA: First floor   816 sq. ft.
Second floor   528 sq. ft.
Heater-Storage   72 sq. ft.
Decks   342 sq. ft.

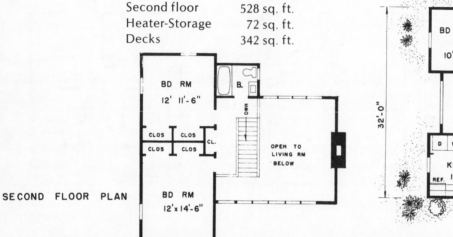

SECOND FLOOR PLAN

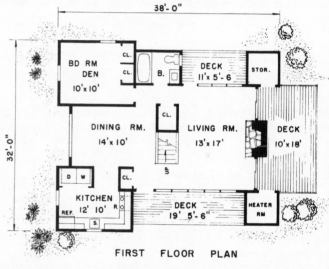

FIRST FLOOR PLAN

The Quebec

Kept to a minimum for the sake of a limited budget, this compact contemporary two bedroom design is a simple rectangle, 24' x 36', featuring a gentle sloping saddle roof taking in a carport on one side and a porch on the other.

The indoor-outdoor character of the plan is increased by the glass wall and door between the living-dining area and the porch.

For year-round living, provision is made for heating and/or cooling unit in the utility room by means of ductwork or, if desired, by electric heating coils in the ceiling, baseboard radiation or wall units.

For the growing number of families who desire to own a retirement or "minimum" home, there isn't much doubt about the suitability of this design.

AREA: First floor   864 sq. ft.
Porch   240 sq. ft.
Carport & Storage   350 sq. ft.

The Brookes

Are you looking for a house suitable to a hillside or mountain lake? Look no further, for here is the ideal house, incorporating substantial size and solidity with the informality of a vacation hideaway.

Two bedrooms, cheery efficient kitchen with easy access to dining area and outdoor patio combined with front to back living room complete with oversized fireplace and balcony suggest the varied use of this house. Downstairs is the perfect open room for children's play or family parties accessible to the lower outside area for picnics and family gatherings.

TOTAL LIVING AREA: 855 sq. ft.

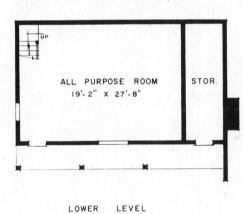

LOWER LEVEL

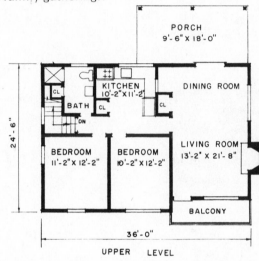

UPPER LEVEL

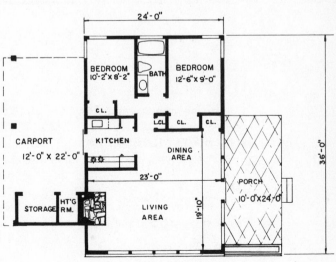

FIRST FLOOR PLAN

The Valley-View

The prow-Alpine roof of this A-frame design is enhanced by the triangular gable and upper deck over the carport. Open planning is stressed by the one large visual area which serves as the kitchen-dinette area and features a stone faced fireplace in the living room and the bedroom on the ground level.

The two bedrooms, each with its own deck, and bath on the second floor can be built at a later date, if so desired. For a high degree of livability and minimum maintenance requirements, there isn't much doubt about the suitability of this A-frame design.

AREA: First floor   868 sq. ft.
        Second floor   505 sq. ft.
        Covered & Open Patio   469 sq. ft.

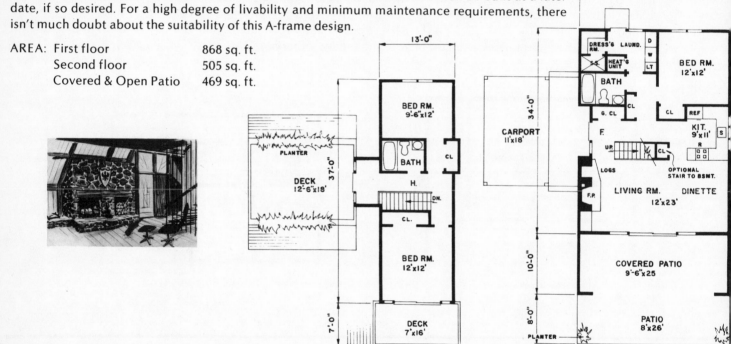

The Pocono

There is an unmistakable Provincial air to this unique leisure home.

The soaring "mansard" type roof is the dominant exterior feature of this design; it is enhanced by the massive fieldstone chimney, vertical V-joint red cedar siding, a sundeck that provides ample outdoor living space and stained wood roof shingles that comprise most of the sides of the building. For low maintenance costs, the interior throughout, except the bath, is of wood paneled walls.

To the growing number of families who today desire to own a "minimum" or second home, there isn't much doubt about the suitability of this design.

AREA: First floor   884 sq. ft.
        Second floor   442 sq. ft.
        Sundeck   364 sq. ft.

The Hide-away

Whether the setting reflects the majestic beauty of a winter scene or the tranquil splendor of a summer landscape, this A-frame design fills the bill for fun-time vacation or year round informal family living.

AREA: First floor   884 sq. ft.
       Second floor   441 sq. ft.

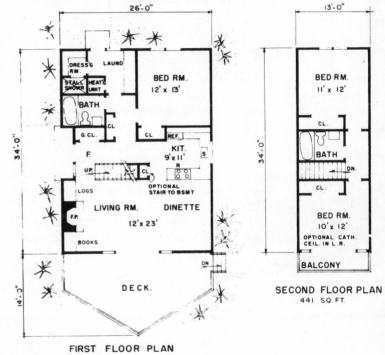

FIRST FLOOR PLAN

SECOND FLOOR PLAN
441 SQ. FT.

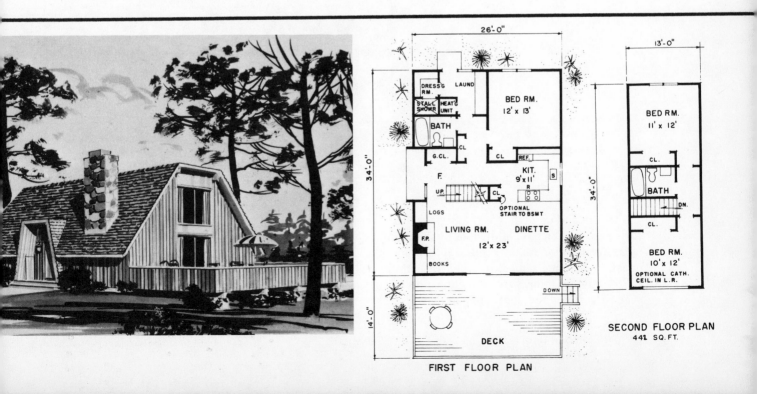

FIRST FLOOR PLAN

SECOND FLOOR PLAN
441 SQ. FT.

The Yukon

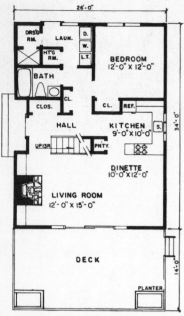

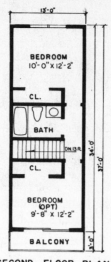

SECOND FLOOR PLAN

FIRST FLOOR PLAN

It is no wonder that the unique look of an A-Frame has proven to be a popular vacation des[...] since it is dramatic to look at, practical to live in and economical to build. The natural eart[...] feeling of this home would be ideal for a wooded or seaside lot in any neighborhood.

Highlights of this design are the fieldstone chimney that soars up through the roof, vert[...] boards and battens, stained red-cedar wood shingles and a red-wood sundeck that creates an[...] teresting exterior.

For year-round living, provision is made for a supplemental heating unit in the utility room. [...] though the plan is of basementless design, a full basement is possible if the physical land c[...] acteristics permit, with the basement stair located under the main stair where the closet is [...] shown.

There is no doubt that this plan typifies the trend toward year-round use of vacation living a[...] leisure activity.

AREA: First floor   884 sq. ft.
       Second floor   441 sq. ft.
       Deck   364 sq. ft.

The Alpine

Modern and up-to-date is the best way to describe this design of a "second home" or a home for all seasons, which is highlighted by the dramatic split roofline, the exterior vertical siding and the massive interior stone-faced fireplace and chimney that rise majestically from the living area past the balcony, clerestory windows and through the shed roof.

Strictly contemporary in feeling, this leisure home is so basically right and appealing that it seems destined for indefinite popularity.

AREA: First floor   960 sq. ft.
       Second floor   580 sq. ft.
       Wood deck   460 sq. ft.

The Lake-Edge

In this distinctive variation of the popular "A-frame," the structure is built in the conventional manner of wood studs, rafters and joists. The entire front facade is glass, making it a wonderful house for a view-endowed property, permitting sunshine to stream into the living area to create a cheerful outdoor atmosphere.

Although the plan is of basementless design, a full basement is possible with the basement stair located where the utility room is now located.

AREA: First level   962 sq. ft.
       Second level   578 sq. ft.
       Total   1,540 sq. ft.

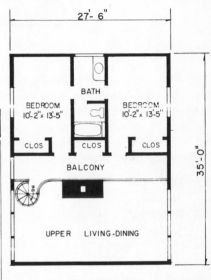

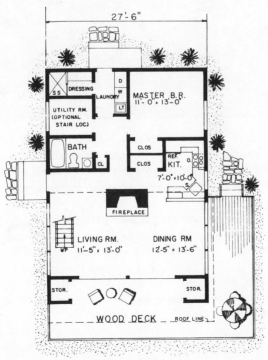

FIRST FLOOR

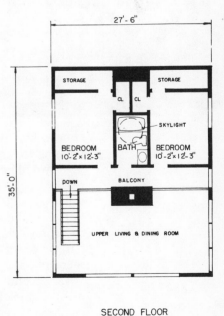

SECOND FLOOR

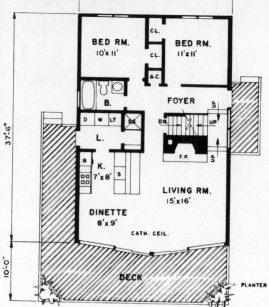

26'-0"

BED RM.
10'x11'

BED RM.
11'x11'

CL.

CL.

G.C.

B.

FOYER

DN.

UP

F.P.

37'-6"

D W LT SK

L.

R K.
7'x8' S

DINETTE
8'x9'

LIVING RM.
15'x16'

CATH. CEIL.

10'-0"

DECK

PLANTER

13'-0"

2ND
FLOOR

BED RM.
12'x13'

B.

CL.

DN.

CL.

CRAWL SPACE

STORAGE

HEAT'R
RM.

GARAGE
11'x25'-6"

F.P.
(OPTIONAL)

REC. RM.
12'-6"x14'-6"

Brookside

There isn't much doubt about the popularity of "A" frame houses in vacation areas. . . . This one permits the opportunity to have one with its own individuality. . . . The kitchen is self-contained and easily handled at the bar, or best of all, when weather permits, outside on the sweeping deck. . . . A fireplace in the living room is a delightful country estate extra. . . . There is plenty of room in this design for indoor as well as outdoor relaxing or entertaining. . . . On the lower level, the recreation room offers added activity, and the garage can be used as an alternate boat stall.

AREA: First floor   975 sq. ft.
       Second floor   316 sq. ft.
       Basement   255 sq. ft.
       Garage   280 sq. ft.

The Beach Haven

The up-to-date modified contemporary exterior styling of this two-story design offers an eye-pleasing effect, which compliments the proportions of the dramatic interior layout.

Redwood boards and battens, striking roof lines, large unobstructed glass area in the front and rear of the living-dining area and the wrap-around wide sun deck give a choice of locations for sunning and relaxing.

The combination living and dining area is most impressive with the generous use of glass, its cathedral ceiling, double pair of sliding doors on the side, cozy fireplace and the charm and intrigue of the overhanging balcony. A dramatic set of wrought iron spiral stairs lead up to the second floor.

The second floor, which may be finished at a later date, consists of two bedrooms, with twin-beds or bunk-house sleeping arrangement, ample closet space, and a connecting bath with mechanical ventilation and ceiling skylight.

This distinctive design is geared for the comfortable seclusion of couples or small families to enjoy carefree year-round living with all the conveniences found in homes costing much more.

AREA: First floor   998 sq. ft.
       Second floor   548 sq. ft.
       Sun Deck   550 sq. ft.

220

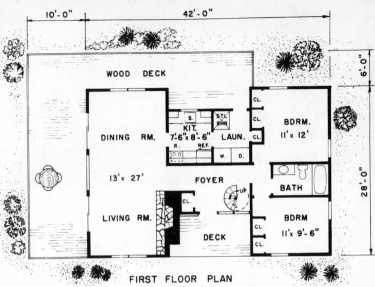

FIRST FLOOR PLAN

10'-0" 42'-0"

WOOD DECK

DINING RM.

13' x 27'

KIT. 7'6" x 8'-6"

LIVING RM.

FOYER

DECK

LAUN.

S. SHR

STL SHR

CL.

CL.

CL.

BDRM. 11' x 12'

BATH

BDRM. 11' x 9'-6"

6'-0"

28'-0"

The Delaware

Today, more than ever before, American families are investing in the future, and one of the most attractive investments is a second home. This "away-from-home" modified chalet-style design offers an abundance of appealing features for indoor or outdoor easy and relaxed living.

The isolated location of the second floor "loft" suggests its use as a guest room or a place for painting or hobby; it features two closets, a balcony that overlooks the living room below, and a private outdoor balcony for sunning, sleeping or viewing.

With only 1,028 square feet of livable space on the first floor this house is designed for economy in construction and is well suited for carefree year round living.

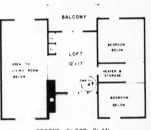

SECOND FLOOR PLAN

BALCONY

LOFT 12' x 17'

OPEN TO LIVING ROOM BELOW

BEDROOM BELOW

HEATER & STORAGE

BEDROOM BELOW

AREA: First floor   1,028 sq. ft.
        Second floor   245 sq. ft.
        Deck   755 sq. ft.

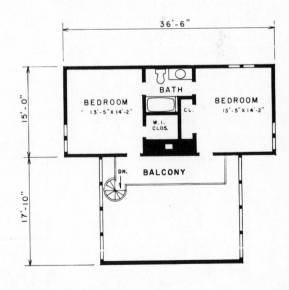

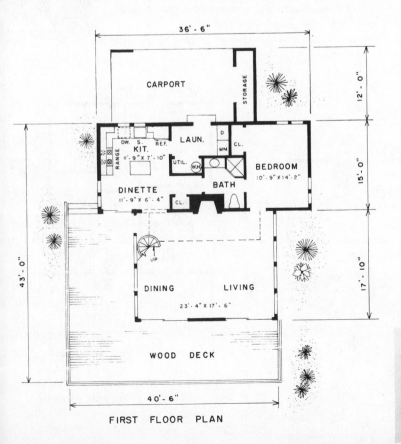

36'-6"

CARPORT

STORAGE

KIT. 11'-9" x 7'-10"

RANGE

DW. S. REF.

LAUN.

UTIL.

WH

D

WM

CL.

BEDROOM 10'-9" x 14'-2"

DINETTE 11'-9" x 6'-4"

BATH

CL.

DINING

LIVING 23'-4" x 17'-6"

UP

WOOD DECK

43'-0"

12'-0"

15'-0"

17'-10"

40'-6"

FIRST FLOOR PLAN

36'-6"

BEDROOM 13'-5" x 14'-2"

BATH

W.I. CLOS.

CL.

BEDROOM 13'-5" x 14'-2"

DN. BALCONY

15'-0"

17'-10"

SECOND FLOOR PLAN

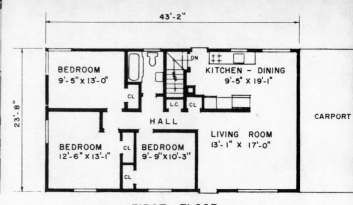

FIRST FLOOR

BEDROOM 9'-5" X 13'-0"

KITCHEN - DINING 9'-5" X 19'-1"

HALL

BEDROOM 12'-6" X 13'-1"

BEDROOM 9'-9" X 10'-3"

LIVING ROOM 13'-1" X 17'-0"

CARPORT

43'-2"

23'-8"

The Rowland

Ranch style economy is the keynote of this condensed version of today's popular home. A full complement of rooms including 3 bedrooms, spacious kitchen dining area and wonderful closet space are important features of this house. Also notice the direct access from the kitchen to outside and to the full cellar. Available with carport, this compact ranch is fully employed to the best use.

TOTAL LIVING AREA: 1,031 sq. ft.

The Seaview

The casual air and the easy care associated with vacation life are assured by the rough-cut vertical siding, stained wood roof shingles, interior wood-paneled walls and the rugged field-stone chimney by the contemporary exterior styling of this design.

Full advantage of the open-plan concept has been taken in the living area by the design treatment of the sunken conversation pit with built-in bench seating facing an open fireplace.

The isolated location of the second-floor loft suggests its use as a guest room, painting, sewing or hobby room. It features two closets and overlooks the living area below.

Although the plan is of basementless design, a full or partial basement is possible if the terrain or physical land characteristics permit, with the stairway from the laundry room.

This plan, with its natural wood exterior and interior, simple design and economical requirements, brings the second home within the reach of many families.

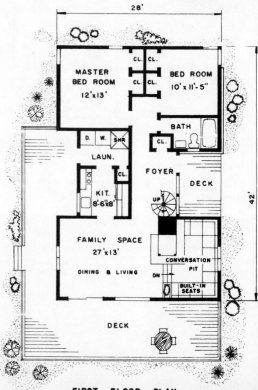

AREA: First floor   1,055 sq. ft.
Second floor   250 sq. ft.
Deck   755 sq. ft.

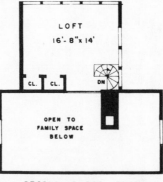

LOFT 16'-8" X 14'

OPEN TO FAMILY SPACE BELOW

SECOND FLOOR PLAN

28'

42'

MASTER BED ROOM 12'x13'

BED ROOM 10' x 11'-5"

BATH

LAUN.

FOYER

DECK

KIT. 8'-6"x8'

FAMILY SPACE 27'x13'

DINING & LIVING

CONVERSATION PIT

BUILT-IN SEATS

DECK

FIRST FLOOR PLAN

The Alden

A great deal of living is designed into this compact ranch plan.

Ideal for a lake or beach home, this may well fit on a suburban lot.

The clean sweeping contemporary lines of the exterior architecture foretell the free open living pattern to be established within.

A full basement is provided in this plan, and there is ample space for play room, utilities, workshop and many other facilities.

TOTAL LIVING AREA: 1,100 sq. ft.
(excl. breezeway and garage).

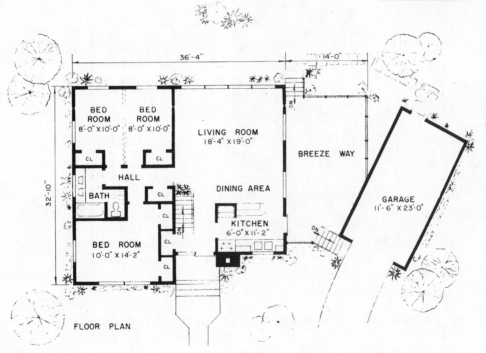

FLOOR PLAN

The Margate

This plan is at home with the great outdoors—it is designed to blend with the landscape and take full advantage of a beautiful view. The natural wood exterior of vertical boards of redwood, cedar or pine, the wood paneled interior, and the other economical construction requirements of this type of plan bring a second home within the reach of many families.

The exposed ceiling wood beams of the dinette-living area slope with the angle of the roof. This area runs the full width of the structure and takes full advantage of the unrestricted view through the all-glass facade and features an attractive log-burning fieldstone fireplace that takes the chill off cool nights. Two bedrooms with ample closets and wall space complete the layout.

There is only 1,040 square feet of livable area and 510 square feet of deck in this comfortable plan, and all of it is well planned and convenient to use.

TOTAL LIVING AREA:	1,040 sq. ft.
Deck	510 sq. ft.

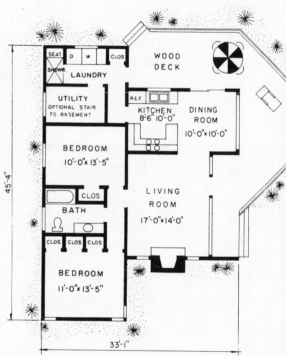

The Harvey Cedars

This contemporary two bedroom ranch design is ideal for retirement—not so large that it would burden a housekeeper, but with plenty of room when the grandchildren come to visit, and excellent for a second home—a lakeside or mountain retreat for all seasons.

The living room is accented by the massive fieldstone fireplace flanked on both sides from floor to ceiling with vertical windows and glass panels, cathedral ceiling and a full wall of windows to take advantage of a good near or distant view.

If convenient, economical and comfortable living is of primary importance, this contemporary ranch-style design that takes advantage of surrounding scenery in almost any direction may be just the new home for you.

AREA:	First floor	1,105 sq. ft.
	Deck	476 sq. ft.

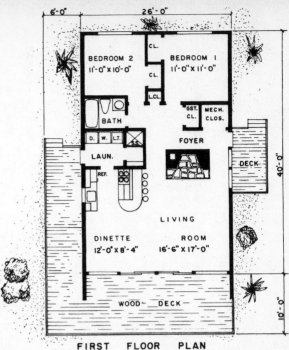

FIRST FLOOR PLAN

The Tahoe

Open planning is stressed by the interior of this design and it is kept rustic with exposed beams at the ceilings and wood paneled walls throughout.

The combination of living-dining area, 30' long, is most impressive with the generous use of glass, its cathedral ceiling, sliding-glass doors to the outdoor deck and cozy fireplace. Ample opportunity is offered by the wrap-around sun-deck to spend hours in the fresh air and make outdoor living, entertaining and serving a pleasurable event.

A dramatic wrought-iron spiral stairs leads upstairs to the balcony bedroom or "sleeping-loft" that may be used by a guest or as a place for painting or hobby.

Whatever your leisure activity, the specifics of this plan are "decked-out" for economy, convenient living and complete relaxation.

AREA: First floor   1,183 sq. ft.
        Second floor loft   210 sq. ft.
        Sun Deck   490 sq. ft.

LOFT PLAN

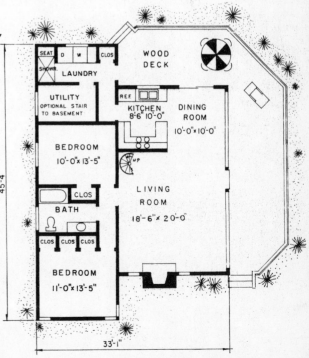

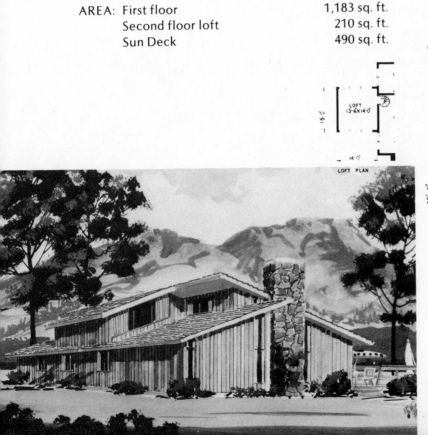

DOME HOMES

Introduced about 20 years ago, this unique concept of living is today enjoying new phenomenal popularity.

Technically, the dome home originated from the sphere, nature's most favored and efficient means of enclosing unobstructed floor space economically.

Due to inflation and the continued ever increasing building construction costs and the fact that the factory assembled triangular space frames are simply bolted together on the site to form the finished building, drastic reductions are possible on quantities of building materials and on-site labor costs. As much as 20 percent less for a dome home than for conventional housing.

The dome provides a living area that also answers the need for efficient energy consumption and is particularly adaptable to solar heating.

Today's dome homes are attractive and offer an exciting new way of living. The minute you step inside the front entrance you are surrounded by fascinating forms and deceptively large spaces.

Because air naturally travels in a circular pattern, heating and/or cooling a dome home is more efficient and economical.

The dome provides maximum enclosed space with minimum surface area which means efficiency in terms of heat-gain or heat-loss. It has been estimated that heating and cooling costs can be reduced by at least 25 percent.

Americans are shedding the conservative trappings of their urban life and are adopting a more youthful, modern and exciting life style.

The two dome home designs on the following pages will be of great interest to you . . .

The Pentagon

This design in today's "dome homes" is among the best bets in efficient and economical housing. With the den used as an alternate second bedroom, the step saver kitchen, full bath and spacious living-dining area—you have all the space you need for a year-round residence, a vacation home in the mountains, on a lake, or at the ocean. If you desire—the second level may be omitted to save construction costs and increase the marvelous feeling of dome space living.

AREA: Ground Floor   1,085 sq. ft.
       Loft   380 sq. ft.

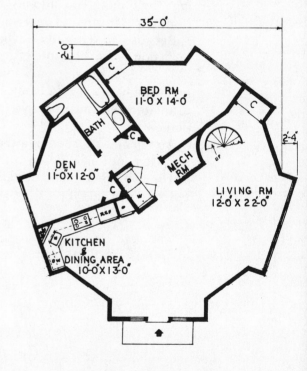

GROUND FLOOR PLAN

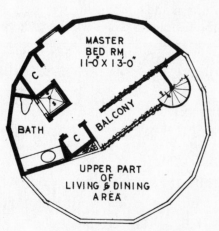

LOFT PLAN

228

The Leisure-Dome

If you have not stepped into the inside of a dome home, you are in for a real surprise.

Today's dome homes are attractive and offer an exciting new way of living. The minute you step inside the front entrance, you are surrounded by fascinating forms and deceptively large spaces.

In this design the dominant exterior architectural feature is the butterfly-roofed porte cochere that is screened by the decorative fieldstone wall and water fountains; but, by far the most dramatic space in the home is the panoramic living-dining-balcony space topped off by gigantic pentagon skylights that let the starlight in and show the clouds drifting by.

A decorative metal circular staircase provides ready access to the upper balcony that overlooks the living area below and leads to the master bedroom suite.

If an exciting new way of life is of primary importance, make a comfortable transition from a flat-ceiling, vertical wall home to a dome type residence.

AREA: Ground Floor   1,135 sq. ft.
        Loft   385 sq. ft.
        Storage Area   56 sq. ft.

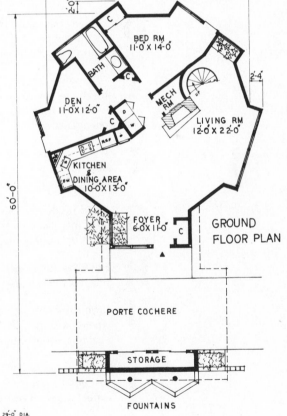

GROUND FLOOR PLAN

LOFT PLAN

229

A HOUSE PLANNING GUIDE
FROM A TO Z

A After you have chosen your building site, obtain the services of a land surveyor to provide you with a topographical survey of your property. The survey should include grade contours, lot lines (their direction and length), location and depth of sewers (if available), water main, gas, electric, etc. All easements, existing trees and other physical property characteristics should be clearly indicated.

B Before beginning preliminary sketches, it is recommended, that copies of all rules and regulations governing the building activity of your area be obtained. This includes a local building code, local zoning restrictions, fire underwriters regulations, local, city or state sanitary requirements, etc.

C Committing yourself to a construction contract for the erection of your home is a matter of great and serious concern. If you do not have the cash necessary to pay for the entire construction cost, a building loan will be needed. A building loan or mortgage may be obtained from your local bank, building loan companies, savings and loan associations, insurance companies, mortgage firms or private individuals. A long term amortizing mortgage with monthly installments arranged like rent is the most convenient. These installments include interest, insurance, payment on principal and frequently taxes and water.

D Design your house to be in harmony with those in the neighborhood. Strive for architectural appeal by simple lines that will lend dignity to the structure. A well designed house gets a high mortgage rating.

E Economy in placing one bath over or adjacent to another is desirable, but this calls for discretion. An apparent saving of say $100. in plumbing might be more than offset in square foot loss and could inconvenience the circulation or convenience of the occupants.

F For an accurate estimate of the cost of your home, submit your plans and specifications to a builder or contractor. Cost per square foot is a good "rule of the thumb" figure but may vary depending on special built-in features, building codes, etc.

G Good residential design requires sound imagination, thought, originality and experience that can only be obtained from an architect. If you intend to use stock plans to construct your home, be sure they are the work of an architect, not a designer.

H Have the title searched on your property. This will protect you as the owner. When the title has been cleared you will get a deed which should be recorded in the proper court.

I If you wish to maintain the value of your property through the years, select a lot where the zoning ordinances have been established. They will protect your property against the encroachments of business, rooming houses, multi-family dwellings and other adverse influences.

J Just build a home to meet your immediate needs; don't go for too big a house. Don't expect to get all your ideas in one house but decide on a plan that is a compromise.

K Keep an open mind on new materials and methods. Consider building in several stages; what you need, build now; what you might need, build later.

L Land is scarce and getting more so every day. You will pay more for land today than it cost several years ago and chances are that it will cost more in the future. When you have decided on the lot make sure you obtain an owner's title insurance policy.

M Modern residential building construction is a complicated job in which scheduling of the work of various sub-contractors is very important. Masons, plumbers, electricians, plasterers, tile setters, carpenters, etc., must be coordinated with each other to avoid time- consuming costly delays.

N Neighborhoods with houses of different syles and prices is good. Stay clear from areas that seem to have all the houses built from the same basic design.

O Once you have determined your own particular home requirements, (two, three or four bedrooms) the only other cost factors are the size of house in square feet, number of baths and lavatories and the amount of livability per square foot.

P Plan your house to fit the lot and avoid the costly need of changing the existing topography to fit the house.

Q Quite apart from the topography of your individual lot you should note the terrain of the surrounding land. Proper drainage of your lot as well as the adjacent property is of utmost importance.

R Recommendations on good building contractors should be obtained from your architect, lawyer, real estate broker, lumber dealer, building supply house, bank or other lending institutions. Ask these builders for a list of the houses they have recently completed and it may be worth your while to visit and talk with some of the people whose homes they built.

S Select a local reputable contractor or builder, and make only written agreements in order to avoid future misunderstandings.

T Trick designs adversely affect the value of your property. It is unwise to build a home that is radically different in an "established" neighborhood.

U Unless you have unlimited funds and can afford to experiment and make mistakes, do not accept radical architectural designs, untried new materials and mechanical equipment. Unless you have a background in the building field, don't count on saving money by trying to act as your own general contractor.

V Visit your lot several times on different days and under different weather conditions. Carefully check the surrounding neighborhood and the orientation of the lot. A pleasant view adds to the enjoyment of life; a viewless lot, however, can be greatly improved by anyone with a talent for landscaping and gardening.

W When you apply for a loan, the bank or other lending institutions will want to know exactly where and what you intend to build. Bring along a copy of your house plans and specifications, a plot plan and a legal description of your property. Be sure you are prepared to establish your financial responsibility.

X Extraordinary precautions should be exercised in establishing your cost limitations. Your home may represent THE largest investment you will make in your lifetime. There is no substitute for good planning, good materials, good workmanship and safe and sound financing.

Y Your first step in building your home should be to consult a lawyer. His fee is moderate and his services, priceless. Explain your proposed building program to him. He will advise you about local procedures and will protect you from making costly mistakes every time you sign your name to a contract agreement.

Z Zoning regulations control what you can build on your lot and are a definite protection for the homeowner, because they keep commercial and industrial neighbors out of residential areas and thus tend to hold up real estate and property values.

WHICH HOUSE FOR YOU, CONSTRUCTION COSTS AND MORTGAGE FINANCING

Once you have determined your requirements and made up your mind to build your new home, further decisions will be much easier to make, if you familiarize yourself with the basic type or design, personal preference and budget patterns. Whether the design you select is a ranch, split level or two story, the descriptive title should not confuse you with the architectural styles such as Colonial, French, English, farmhouse, etc. which may be adapted to any one of them.

The debate is never-ending on the relative merits of ranch, split level and two-story homes.

Evidence can be marshalled by advocates of each style that their favorite is best for all around economy, livability, and other virtues.

Rectangular perimeters offer the least expensive base construction, and simple straight roof lines the most economical cover for the base. Starting with the least expensive, a small rectangular one story building at a fixed area and price; a two story house with the same size foundation and roof would give double the living area but less than double the cost. A split-level of the same original foundation size would increase the living area over the one story. This is usually accomplished by "lifting" the bedroom area so that the basement floor below comes up to grade level thereby providing additional living space on grade in cellar space which had been previously below ground. This "lifting" process causes some extra expense in framing and roofing, but provides the extra living area at less cost than it would to obtain it in the original one story building by making a larger foundation area, and shaping the ranch into a familiar rambling "L," "U" or "T" shape.

The debate cannot be resolved on construction cost alone. There are other features desirable and important in the arrangement of the plan to afford comfortable and convenient living. One point which all agree on—there should be definite separation between living, sleeping and recreation areas. Here also a controversy arises— should this separation be horizontal or vertical? The ranch house can effectively provide this separation by proper planning of the interior layout. This is more difficult with the simple basic rectangle—hence the rambling feature usually present in the ranch. It is simple to design a bedroom wing, and a recreation wing if they can extend in any direction away from the central core of living area. When stairs are no objection the two story layout provides this separation quite definitely and most satisfactorily. In the split level or other multi-level arrangements there are usually 3 basic living levels and each is quite positively separated by a short flight of stairs thereby carrying to a finer degree the separation between living, sleeping and recreation areas by providing a separate level for each.

In this modern age the sleek streamlined look seems most appealing to many people even though they strive to obtain the character of detail inherent in the old colonial architecture. The two-story house, to obtain this effect would have to be large and generally will give a more stately rather than streamlined appearance. The ranch or one story home can have this pleasing effect even though small. If it is larger and is designed in some shape other than rectangular, the "rambling ranch" appearance is very attractive.

Split levels are highly adaptable to many styles of exterior appearance. Ingenious arrangements of roof line and adjustments of levels can give two-story as well as ranch-like character to the conventional split level home. The front-to-back and back-to-front split levels can even more strongly suggest a ranch type home in the former case and a two-story in the latter. Purely for its own style, the conventional split-level suggests a house of separate wings, each for its own specific use generally considered to be a sign of an expensive home.

Economy in heating and plumbing is another

feature strongly debated by advocates of the different types of homes. The lower level recreation area of the split-level has long been a thorn in its side when it comes to proper heat. The introduction of the 2 zone system has eliminated this to some extent when using forced warm air. The usual concrete slab floor construction here has some advantages when using hot water heat however. The installation of hot water piping in this concrete and the resultant warm floor and radiant heat provide the most comfortable area in the house in winter.

Proper insulation of course is the best controlling factor for economy of operation in any heating system. It will protect a bedroom floor when it is located over an unheated garage or a wall of a room located next to a garage. Of course exterior walls and ceilings at unheated attic areas are "must" locations for insulation. Most economical plumbing arrangements concentrate in one small area those rooms that require plumbing. Small homes have baths located next to the kitchen with a result of less separation between sleeping and living areas. Ingenious hall arrangement can overcome this proximity somewhat but for complete separation in the ranch plan separate plumbing areas can be expected to increase costs.

HOME FINANCING: Owning your own home has been greatly simplified during the past thirty years. The modern long-term, low interest self-amortizing mortgage, (covering principal, interest, taxes and insurance) has changed the whole institution of home buying. Under this plan the prospective home owner makes regular monthly payments on the money that is borrowed. The Real Estate taxes and the interest you pay on your home are income-tax deductible.

There are basically three different types of mortgage loans; conventional, FHA and VA; (U.S. Housing & Urban Development).

Conventional loans are usually obtained through the banks and other institutions. Since there is just so much money to lend these days, they are trying to make the best of a bad situation, by getting more and more selective at higher interest rates, than FHA or VA loans.

National, State and Mutual Savings Banks, Building and Loan Associations, Insurance Companies, and Mortgage Bankers represent the major mortgage lending agencies. In the long run most people find dealing with their own local bank or institution far more satisfactory than dealing with an agent representing other institutions.

The Federal Housing Administration (H.U.D.) does not make mortgage loans. Instead it insures the mortgage loans made through some 30,000 lending agencies. This protection enables lending institutions to make insured mortgage loans on desirable terms, with a small down payment and government-limited financing rates.

The Veterans Administration also guarantees GI loans through regular lending institutions. If you are an ex-serviceman, or woman, it may qualify you for a lower down payment and a longer term mortgage than civilians who have not served the armed forces are eligible for.

From time to time the government varies both the percentage of down payment required, the maximum number of years the mortgage may run, and the prevailing interest rate. Check with your local lending institution to see what rates are current before you start.

CONSTRUCTION COSTS: There are several methods of estimating the approximate costs of any new home. The one most used by Architects, builders and appraisers is the square foot method. Geographic locations vary the cost of both material and labor. Local building conditions and codes differ to such a wide degree that an accurate unit scale is almost impractical. Generally speaking, construction costs range from $25 a square foot of living area, assuming that the work is contracted out to a Contractor. Any work that you may do yourself such as painting, decorating, landscaping, etc., would reduce the cost.

Remember, that only your builder can give you an exact and final building cost figure, and that the rule-of-thumb yardstick, as out-lined above, is merely for your generalized fireside consultation. By multiplying the square foot area by the construction estimate, you will be able to catalogue the design that interests you most into a general price category. (The cost of land, of course, will be entirely separate.)

SELECTING A BUILDER: To build the home you have selected requires the services of a reliable contractor. Recommendations may come from friends who have built or are building a new home, or perhaps you can obtain the names of the contractors who may be constructing homes in the newly developed areas of your town.

If you know someone who had a home built and was satisfied with the result, ask for the name of the contractor.

Since you will definitely want to obtain several bids, interview several contractors, and if possible, visit some of the homes that they have built during the last few years.

Many builders belong to the National Association of Home Builders and although the NAHB is a national organization which officially credits home builders as to a certain level of professionalism; a small builder,—one who builds say, (two or three houses a year) may not belong, but still be competent and reliable.

And finally, do not sign any papers or agreements without the presence of a lawyer's services who could help you avoid extremely costly mistakes in dealing with the builder, title company, or money lending institution.

HOW TO BUILD YOUR HOME

Your first step in building your home should be to consult a local lawyer. His fee will be moderate and his service priceless.

Both husband and wife should attend when the lawyer is involved in discussions and paper signing. Explain your proposed building program to him. He will advise you about local procedures; he will protect you from making costly mistakes and he will be on hand every time you sign your name to a contract or agreement.

BUY PROPERTY

In most cases, you will be unable to obtain mortgage financing without ownership of the property on which your house is to be built. So this is the next step.

When you have found a lot that meets all your requirements, call in your lawyer. He will determine whether you really will be the owner of the land you are ready to pay for. The seller must be able to furnish you with a "clear title." Your lawyer will advise you how to proceed on this.

While this title search is going on, a prudent way to protect your interests is to have the deal held in escrow. That means turning the purchase price over to a third party (your own or the seller's lawyer; the real estate broker; your bank or title company) until the title is cleared. Once the deal is in escrow, you can proceed with the plot survey. Engage a local surveyor or civil engineer because he probably has done other work in the neighborhood and has time-saving data on file in his office, which will be reflected in his fee.

A complete plot survey shows on paper every outline, every angle, every dimension of your plot. The location, size and depth of underground sewers, water mains and gas lines should be plotted with the house connection stubs, if any. The survey should show the location of adjacent houses; if any, nearest your line on either side to permit placing of your house to secure maximum privacy, light and prevailing breezes.

A plot survey includes permanent markers on the ground at every corner and at every angle if the plot is irregular.

While your title is being searched, arrange to take out a title guarantee policy. It usually is cheaper to get an owner's title insurance policy from the company making the title search in connection with the sale. This is because the search and examination will not have to be duplicated and the cost of this loss-prevention work on the part of the company accounts for the bulk of the title policy charge.

The big safeguard in title insurance to you is that the title company must defend any claim made against your ownership. The cost of such a defense could exceed the cost of your whole home. The fact that the mortgage lender will also carry title insurance is not adequate for you; his covers for the amount of the mortgage; your title insurance must also protect your equity over the amount of the mortgage.

When your title has been cleared you will get a deed. Have this recorded in the proper court. You will pay a revenue stamp tax on the purchase price.

You have now acquired the site. It is protected against trouble. You are now ready to build your new home.

ARRANGING A LOAN

Rarely do families have the amount of cash necessary to pay the entire construction cost of the home; you will probably need to borrow money to build. What you need is a building loan.

This building loan is usually converted automatically into a mortgage when the house is completed. Terms of the mortgage will be established when you arrange the building loan.

Usually a builder will not start work without some down payment and an agreement on a schedule of payments to be made at regular intervals during the course of construction. Find out the financial requirements of the builder you select and establish if he is to be paid directly by you or the lending institution. Your lawyer will help you. As a guide for your reference, here is a typical schedule of payments to the contractor while the house is being built.

10% on completion of foundation

25% on completion of the rough enclosure

30% on completion of the plastering, plumbing, heating and electricity

25% on completion of the work

10% 30 days after completion of all work.

Withholding of the final payment for 30 days is to insure correction of any defects or oversights. This should be mentioned specifically in your agreement with the builder which your lawyer will draw up for you.

At the time each payment is made to the builder, have your lawyer make certain that it is in accord with the original agreement and that you receive a proper statement of receipt. Before the final payment is made your lawyer should carefully verify that there are no liens or outstanding unpaid bills that might become a claim against you.

You can get your building loan and mortgage money from banks, local building and loan companies, or mortgage firms.

These lending institutions are in business to make money and they have just one commodity for sale—money. There is real competition among lending agencies for your business. Do not hesitate to shop around for terms.

Talk to a number of these representatives but do not make out a formal application for a loan until you have studied their offers. Most homeowners find the long term mortgage, with monthly installments arranged like rent, the most convenient. These installments include interest, insurance, payment on principal and, frequently, tax and water charges.

Interest rates vary. A fraction of 1% saved each year amounts to a sizable sum over the term of your mortgage. Depending on conditions in the money market, rates vary. Even though you may not need the biggest loan you can get, it is assuring to know you could raise more funds.

Check to see how much the cost will be, if at some later time you might wish to pay off your mortgage because of a gift or inheritance, and after what period you can repay without penalty.

When you apply for a loan, the bank or other lending institution will want to know exactly where and what you intend to build. Be sure you take with you a copy of the house plans and specifications, and plot plan or short legal description of the property.

You must be prepared to establish your financial responsibility as a good risk. This means a statement of your assets and liabilities, income and employment record. A good rule of thumb is that 20% to 25% of your yearly gross income should equal or exceed your yearly payments on principal, interest, taxes, insurance and maintenance of your home.

Here are some questions from a typical loan application form: What is your employment record, position held, salary, number of years on job, previous positions with other firms, bank accounts, life insurance and amount of annual premiums, previous mortgage experience, stocks and securities held, other income, number of dependents, judgments or garnishees against your salary?

BUILDING PERMIT

A building permit is generally required before construction is started. Your builder may handle this for you or you may apply for it through your local building department. Two sets of house and plot plans are usually sufficient to submit with the application. One set will be returned with the building permit. A small fee is generally charged for the building permit.

MORTGAGE PAYMENT TABLE

Amounts shown include monthly payments of interest
and principal but not taxes and insurance.

20-YEAR MORTGAGE

Amount	at 7½%	at 8%	at 8½%	at 9%	at 9½%	at 10%	at 10½%
$16,000	128.96	133.76	137.68	144.00	149.15	154.41	159.75
20,000	161.12	167.20	173.60	180.00	186.43	193.01	199.68
24,000	193.44	200.64	208.32	216.00	223.72	231.61	239.62
28,000	225.68	234.08	243.04	252.00	261.00	270.21	279.55
30,000	241.80	250.80	260.40	270.00	279.64	289.51	299.52
36,000	290.16	300.96	312.48	324.00	335.57	347.41	359.42
40,000	322.40	334.40	347.20	360.00	372.86	386.01	399.36
44,000	354.64	367.84	381.92	396.00	410.14	424.61	439.29
50,000	403.00	418.00	434.00	450.00	466.07	482.52	499.19
54,000	435.24	451.44	468.72	486.00	503.36	521.12	539.13
60,000	483.60	501.60	520.80	540.00	559.28	579.02	599.03
64,000	515.84	535.04	555.52	576.00	596.57	617.62	638.97

25-YEAR MORTGAGE

Amount	at 7½%	at 8%	at 8½%	at 9%	at 9½%	at 10%	at 10½%
$16,000	118.24	123.49	128.84	134.27	139.80	145.40	151.07
20,000	147.80	154.36	161.05	167.84	174.74	181.75	188.84
24,000	177.36	185.24	193.25	201.41	209.69	218.09	226.61
28,000	206.92	216.11	225.46	234.97	244.64	254.44	264.38
30,000	221.70	231.54	241.57	251.76	262.11	272.62	283.26
36,000	266.04	277.85	289.88	302.11	314.54	327.14	339.91
40,000	295.60	308.73	322.09	335.68	349.48	363.49	377.68
44,000	325.16	339.60	354.30	369.25	384.43	399.83	415.44
50,000	369.50	385.91	402.61	419.60	436.85	454.36	472.10
54,000	399.06	416.78	434.82	453.17	471.80	490.70	509.86
60,000	443.39	463.09	483.14	503.52	524.22	545.23	566.51
64,000	472.95	493.96	515.35	537.09	559.17	581.57	604.28

IS NOW THE *TIME* TO BUILD?

Perhaps you are asking yourself such questions as: Are we ready to build now? Can we go ahead this year? Or will next year be better?

For most families facing this decision, the best answer is to do it *now*. Due to our recent inflationary economy, the value of homes has doubled or tripled during the past twenty years, so the odds are:—that due to economic realities, homes will continue to increase in value because of the expected rise in labor and the cost of materials. Any long term investment like a home must rise in value during an inflationary economy. If you postpone the construction of your new home for even as short a time as one year, you well may find that the cost will have risen by as much as ten percent.

More importantly, the decision to go ahead now will enable your family that much sooner to start enjoying the better living and the enhanced security that your new home will provide. Just as building costs rise, so do the total dollar values of homes; you can safely expect the home you build this year to be worth several thousand dollars more than its cost ten years from now. In the meantime, as a homeowner, you will be enjoying special income tax advantages in the fact that your payments for mortgage interest and real estate taxes are fully deductible. Finally, your payments of mortgage principal will each month decrease the amount you owe; by meeting your monthly payments, your equity grows as automatically as if you were making regular deposits in a savings account.

So, for an investment whose value and equity grows, and the psychological motivation of owning your own home;—the time to proceed with arrangements for the construction of your new home is *now*.

ABOUT YOUR HOME . . .
BEFORE YOU WRITE—READ!

We welcome correspondence and are happy to
answer your letters, but why not save yourself time
and effort. Perhaps the answer to your question is here.

Are cost estimates included or can you tell me how much my favorite house will cost to build?

• Construction costs vary so much from one section of the country to another that you will do better to get a set of blueprints of your favorite plan and obtain an estimate locally. Costs range upwards from $25.00 a square foot of living space, assuming that the work is contracted out. Our designs show the square feet of living area and unless otherwise specified, this does not include porches, terraces, garages, etc., since many of these features are optional and, of course, cost less per square foot to build than the main dwelling. With our blueprints you can get actual local cost estimates from builders and arrange financing with a mortgage lending institution.

Will you make plan changes for us?

• In most instances, changes in dimensions, substitution of items, materials, etc., or minor alterations can be done by the contractor during construction. If the house plan calls for wood siding, it can be changed to brick, stone or other materials; only the width of the exterior walls must be adjusted for the difference. We furnish conversion details, otherwise the working drawings for our designs are available only as illustrated. If major changes are involved, you should consider ordering one set of blueprints and have them redrawn by your local architect.

Will you tell me where a particular house has been built so I can look at it?

• During the past twenty years, we have sold many thousands of our plans for homes that have been built throughout the entire country. It is regretted, that our blueprint buyers, seldom give us any information, as to where or when they expect to build. Our design illustrations are accurately drawn perspectives and with the exception of the landscaping, the house will appear exactly as shown.

Will plans meet local building requirements?

• Our plans have been engineered for sound construction, but as long as there are almost as many different building codes as there are communities there are bound to be rare cases of conflict. There is no need for concern however, inasmuch as any suggested changes can usually be done during construction without the necessity of new or revised plans.

Can I get blueprints "in reverse" with the living room, for instance on the left instead of on the right as shown?

• If you find that your favorite house plan would suit you—or your lot—better if it were reversed, we will upon request, send—one of the sets transposed as in a mirror. Even though the lettering and dimensions appear backward, they make a handy reference because they show the house just as it's being built in reverse from the standard blueprints—thereby helping you visualize the home better. For example, if you order five sets of plans, we will send one mirror image, and four in the original position so that you can read the figures and directions easily.

How many sets of blueprints should be ordered?

• The answer can range anywhere from 1 to 8 sets depending upon circumstances. A single set of blueprints of your favorite design is sufficient to study the house in greater detail. If you desire to get cost estimates or planning to build, you may need as many as eight sets of blueprints. A minimum number of five sets is required, one each for: owner, builder, material dealer, building permit and mortgage financing. In many cases, local building departments require two complete sets of blueprints before they will issue a building permit. (Check with your building department.) You may need six sets.

How is the low cost of your blueprints possible?

• If you had complete working drawings especially created by a personal architect the design fee for an individual home may be 8 to 10 percent of the total construction cost, and could range from several hundred dollars up to several thousand, depending on how big and complicated the design is. When you use our *architect-designed* plans, (prepared by and/or under the supervision of professional licensed architects) the cost is spread among other families planning to build the same house in various parts of the country and they are sharing the total costs with you. Our many years of practical home planning experience assure you of a well designed, practical house that will stay younger longer and make you feel proud of owning the home of your dreams . . .

Do you furnish a description of material or material list?

• All of our working drawings are furnished with a suggested description of materials required to construct the house as illustrated. Many contractors and material dealers prefer however to make up their own material list to take full advantage of materials most readily obtainable at best prices locally, thus permitting the substitution of items to satisfy your personal preference.

Plan Orders Mailed Within 24 Hours!

HOW TO ORDER YOUR BLUEPRINTS

If the design you have selected satisfies your requirements, mail the accompanying order blank with your remittance. However, if it is not convenient for you to send a check or money order, merely indicate C.O.D. shipment.

We will make every effort to process and ship each order for blueprints the same day it is received. Because of this, we have deemed it unnecessary to acknowledge receipt of our customers' orders. See order coupon below for the postage and handling charges for surface mail, air mail and foreign mail.

Should time be of the essence, as it sometimes is—

For Immediate Service
Phone (201) 376-3200

Your plans will be shipped C.O.D. Postman will collect all charges, including postage. (No C.O.D. shipments to Canada or foreign countries).

NATIONAL HOME PLANNING SERVICE

37 Mountain Avenue

Springfield, N.J. 07081   Phone Orders (201) 376-3200

PLEASE SEND HOME DESIGN, BUILDING PLAN NAME — THE _____

First set of Plans (if only one is desired)	$69.00	$_____
Each additional set with original order . . . . @	$15.00	$_____
To have Plans Reversed	$ 5.00	$_____
Five (5) sets of Architect's Total Blueprint and Building Package	$95.00	$_____

ADD THE FOLLOWING POSTAGE:

Parcel Post (allow 2 to 3 weeks)	$3.00	$_____
First Class-Air Mail	$4.00	$_____
C.O.D. (U.S. Only)	$5.00	$_____
Canada and Foreign Air Mail	$7.00	$_____
Make payment in U.S. currency to National Home Planning Service	TOTAL AMOUNT	$_____

Prices subject to change without notice!

MAIL ORDER TO:

NAME: _____

STREET: _____

CITY: _____ STATE: _____ ZIP: _____

NATIONAL HOME PLANNING SERVICE

37 Mountain Avenue

Springfield, N.J. 07081   Phone Orders (201) 376-3200

PLEASE SEND HOME DESIGN, BUILDING PLAN NAME — THE _____

First set of Plans (if only one is desired)	$69.00	$_____
Each additional set with original order . . . . @	$15.00	$_____
To have Plans Reversed	$ 5.00	$_____
Five (5) sets of Architect's Total Blueprint and Building Package	$95.00	$_____

ADD THE FOLLOWING POSTAGE:

Parcel Post (allow 2 to 3 weeks)	$3.00	$_____
First Class-Air Mail	$4.00	$_____
C.O.D. (U.S. Only)	$5.00	$_____
Canada and Foreign Air Mail	$7.00	$_____
Make payment in U.S. currency to National Home Planning Service	TOTAL AMOUNT	$_____

Prices subject to change without notice!

MAIL ORDER TO:

NAME: _____

STREET: _____

CITY: _____ STATE: _____ ZIP: _____

NATIONAL HOME PLANNING SERVICE

37 Mountain Avenue

Springfield, N.J. 07081   Phone Orders (201) 376-3200

PLEASE SEND HOME DESIGN, BUILDING PLAN NAME — THE _____

First set of Plans (if only one is desired)	$69.00	$_____
Each additional set with original order . . . . @	$15.00	$_____
To have Plans Reversed	$ 5.00	$_____
Five (5) sets of Architect's Total Blueprint and Building Package	$95.00	$_____

ADD THE FOLLOWING POSTAGE:

Parcel Post (allow 2 to 3 weeks)	$3.00	$_____
First Class-Air Mail	$4.00	$_____
C.O.D. (U.S. Only)	$5.00	$_____
Canada and Foreign Air Mail	$7.00	$_____
Make payment in U.S. currency to National Home Planning Service	TOTAL AMOUNT	$_____

Prices subject to change without notice!

MAIL ORDER TO:

NAME: _____

STREET: _____

CITY: _____ STATE: _____ ZIP: _____